Solar System

Solar System

Nigel Hey

Weidenfeld & Nicolson

To my family – Brian, Jocelyn, Jonathan, David, and Edward, and my mother Margery, who many years ago convinced a small boy that he was a writer

First published in the United Kingdom in 2002
by Weidenfeld & Nicolson

Text copyright © Nigel Hey 2002
Design and layout copyright © Weidenfeld & Nicolson
2002

Design: Grade Design Consultants, London

A CIP catalogue record for this book is available from
the British Library

ISBN 0304359947

Printed and bound in Italy

Weidenfeld & Nicolson
Wellington House
125 Strand
London WC2R 0BB

Contents

	Introduction	6
	Space Talk	9
1	**The Greatest Drama**	**10**
2	**The View From Here**	**36**
	Humans and Machines, Time and Space *Arthur C. Clarke*	58
3	**Technology, Dreams, and Little Green Men**	**60**
	Bases on the Moon, Mars and Beyond *Stephen Braham and Pascal Lee*	65
	Killer Asteroids *David Morrison*	71
4	**Interplanetary Voyagers**	**88**
	Nuclear Propulsion for Human Interplanetary Exploration *Roger X. Lenard*	108
5	**The Billion-Mile Remote Control**	**114**
	Navigating to Neptune *Donald L. Gray*	127
6	**Our Place in Space**	**136**
	How Rare is the Earth? *Donald Brownlee*	142
7	**The Planetary Neighbourhood**	**162**
	The Starward Sisters *Sarah Dunkin*	181
	Life Abundant *David Darling*	192
8	**Twin Giants**	**198**
9	**The Outer Limits**	**228**
	Exploration of Ice Giants *Heidi B. Hammel*	238
	Ices in the Solar System *Duncan Steel*	252
	Visions of the Future *Freeman Dyson*	258
	Glossary	260
	Index	266
	Further Reading	270
	Acknowledgements	272

Introduction

Years ago, when my career as a science and technology writer was just beginning, I published a short book titled *How We Will Explore the Outer Planets*. At the turn of the present century, while the Galileo spacecraft was on its extended tour of Jupiter, and Cassini was on its way to Saturn, I decided to revisit the subject of planetary exploration. This time I would be able to describe the adventure not as one that might happen, but as one that is unfolding every moment of every day.

There are many fine books about astronomy, and planetary exploration, and space technology. However, for the general public, and for those whose careers will someday take them into space studies, impressions of the solar system come from a collusion of many science and technology disciplines. My intent therefore would not stop at describing what we are learning about the Sun and nine planets. It would also explain how we are making those discoveries, and how economic and political considerations sometimes come into play.

The resulting book research took me on imaginary flights around the solar system, wondering on the mysteries of Mars and its abyssally deep Marineris canyons, peering in awe at the sulphurous landscape of Io, skimming the craggy terrain of Miranda, floating among the shimmering rings of Saturn. I built spacecraft in my mind, blasted them into orbit, and ricocheted around the system as if it were a giant pinball machine, watching the planets and moons go by. Sometimes the lights of a New Mexico dusk would remind me of the colourful bands of Jupiter; on scheduled flights the view between cloud layers prompted me to imagine I was reconnoitring the atmosphere of Venus. I nearly jumped out of my chair when I read Bill Broad's description of Earth's deep-sea microbial life in *The Universe Below*: this sort of thing might exist in Europa's oceans!

The solar system is populated with jewels of great wonder. In the past few decades, humans have come to know these orbiting bodies, which range in size from mighty Jupiter to tiny specks of dust, with increasing intimacy. At the same time that our roving robots are probing Mars, other spacecraft are collecting 'stardust' in an attempt to learn how the system began in the first place. We learn more with every bit of data that comes back to Earth as the result of these investigations – and at the same time we learn how much more there is to discover.

When you think about it, humans have been mapping the solar system for a long time. The process started more than two thousand years ago, when cartographers made maps of earthly neighbourhoods and charted the skies for the determination of seasonal change, the guidance of religious activities, and the provision of astrological signs. Close study of individual planets, however, had to wait for the invention of the telescope in the seventeenth century. Telescopic vision became much more acute three decades ago, when new technologies – in particular charge coupled devices and adaptive optics – arrived on the scene. At the same time, space technology was developing very quickly. Instruments that previously had been designed for more conventional telescopes were soon adapted for use on satellites orbiting the Earth and on planet-probing spacecraft.

Something was happening in these decades that escaped the notice of many. Science and technology, including computer science, were working together. One sees expert space-science papers being published by employees of Lockheed Martin, one of the largest American defence contractors. Some engineers become expert computer scientists. In all fields, science enables new technology; new technology enables new science.

This book reflects the interdisciplinary nature of space studies by concentrating first on the technologies we use to map the solar system, then moving to the discoveries that shape our understanding of the nine planets and the uncountable asteroids and comets that accompany them.

This book is also a tribute to people, who regardless of their educational, religious, or economic background, are blessed with the ability to look into the sky and marvel at the greatness of all that is out there. This uniquely human attribute gives us a sense of perspective, a way of balancing the hubbub of thought that comes with the gift of intelligence. It transports us beyond ourselves and our artefacts. It is another way in which we are able to emerge from our self-centred psychological neighbourhoods, to explore a multidimensional realm where self is of no particular significance. It is one path among many through which individual people may comprehend the close community of all life and all humanity, and, with the accession of humility, the rightness of compassion and peace.

It is true that our descendants may want to colonize other planets in order to preserve our species. And practical benefits will no doubt emerge from space exploration and development. But the greatest gift of all is its contribution to human need for wonder, which it provides with an abundance that is literally beyond end. The scientific curiosity that we all share – that part of our minds that dwells upon physical phenomena that may be of no consequence to our material existence – is no longer confined to Planet Earth.

In researching and writing this book I learned many new things, and came to believe in a few ideas that once invoked my scepticism. First, I believe in the exploration of the solar system more than ever before. Second, I believe there is life elsewhere in our solar system, and what we consider to be intelligent life in other solar systems. Third, I believe we will discover at least one other potentially life-friendly planet in the next quarter-century, orbiting some other star. Fourth, I believe that, barring some global catastrophe, humans will establish a base on Mars sometime in the next fifty years.

I also learned from the men and women who contributed essays for the book. As the contributions came in, I recognized the special value of the writers' intense and personal experience with space science and technology. Naturally their writings deal with spaceflight, the solar system, and the cosmos. Importantly, they also reveal the profoundly human aspects of this great adventure, from the excitement of solving the problems of spacecraft that are millions of miles distant, to the self-examination that occurs when considering whether we might someday send robots as our ambassadors to distant star systems. Some of the nobility of the human condition, so often obscured, shines through these words.

I used to wonder, why bother? The other planets and moons are too inhospitable for us ever to visit, let alone colonize. But this is shallow thinking. Our destiny is in space. We will always want to explore new frontiers, even when separated physically by great gulfs of space and time. The need is in our genes. And who knows what wonders we may find out there.

For the present, my greatest hope is that readers will enjoy reading this book as much as I enjoyed writing it. Hang on, and enjoy the ride.

Space Talk

Every speciality has its own shortcuts to communication. Words encapsulate concepts or ideas in brief bursts of sound. They are linguistic shortcuts to understanding, whether in poetry or invective. Some convey extremely simple, direct information; but many new words are quite the contrary, to the extent that they are used mainly by specialists.

At the same time, many technical terms are making their way into ordinary language. In the past sixty years, dictionaries of all languages fattened as technology became more important to our everyday lives. Complex ideas such as 'moving pictures sent and received as radio waves and displayed on a phosphorescent screen' became truncated into new words, like 'television'.

Nobody knew what a transistor was fifty years ago, when the computer age was still in its infancy and robots were little more than science-fiction creations. Today we have gigs and megs, woofers and tweeters: indeed thousands of words that arrived in our vocabularies courtesy of the information and entertainment marketplaces.

A fair number of people would understand if told about a computer that has a processor that runs at 2.4 GHz, along with a CD-RW/DVD combo drive. This shorthand techno-speak at least conveys the idea of a fast machine that is able to record and play compact disks as well as play pre-recorded video disks. We're not nearly so familiar with the language of space. If there were a report that 'Cassini is having trouble with its downlink to the DSN', relatively few would realize that a spacecraft that is on its way to Saturn is having a problem sending information to Earth.

In the special language of planetary scientists, a day – the time required for a planet to rotate once on its axis – is termed the sidereal rotation period. A year, the time it takes to make one circuit around the Sun, is the sidereal orbit period. The difference observed when the planet's orbital path is compared with the apparent orbit of the Sun around the celestial sphere (the ecliptic) is the 'orbital inclination to the ecliptic'. Mean orbit velocity, orbit eccentricity, and escape velocity are other important terms in the space scientist's lexicon, along with the more obvious characteristics of planetary bodies – distance from the sun, equatorial radius, volume, mass, density, and surface gravity. Knowledge of all these factors, and many more, is important in resolving questions that range from planning a spacecraft trajectory to calculating the amount of solar energy received by each body in the solar system.

A list of technical terms is included in the glossary at the end of this book.

The Greatest Drama

'For a million years it was clear to everyone that there were no other places than the Earth. Then in the last tenth of a percent of the lifetime of our species, in the instant between Aristarchus and ourselves, we reluctantly noticed that we were not the centre and purpose of the Universe, but rather lived on a tiny and fragile world lost in immensity and eternity, drifting in a great cosmic ocean dotted here and there with a hundred billion galaxies and a billion trillion stars. We have bravely tested the waters and have found the ocean to our liking, resonant with our nature. Something in us recognizes the Cosmos as home.'

Carl Sagan, Cosmos. New York: Random House, 1980

Each of us is part of an endless drama that started billions of years ago, when a gargantuan cloud of cosmic gas and dust began to collect within the firmament. Then our galaxy, the Milky Way, came into being, and, within it, our solar system. The Sun is the great nucleus of this little cell, radiating incredible amounts of energy to its daughter planets, furnishing all the ingredients of life and all the analogues of life that manifest themselves as simple movement and change.

The story of the planets starts with the formation of the star that we call our Sun, an awesome, intricately balanced system for the conversion of matter to energy. At its centre is a nuclear furnace that changes hydrogen into helium, electromagnetic energy, and neutrinos. This energy pours continuously outward into the solar atmosphere and a brilliant corona that envelops the entire solar system.

Through the millennia, people have stood in awe of the Sun, revered and worshipped it – from those who gathered at Neolithic Stonehenge to certain modern-day tribes whose beliefs have withstood transformation by Western scientific thought. Modern skywatchers experience at least as much wonder as they witness distant scenes with the aid of spacecraft instruments and

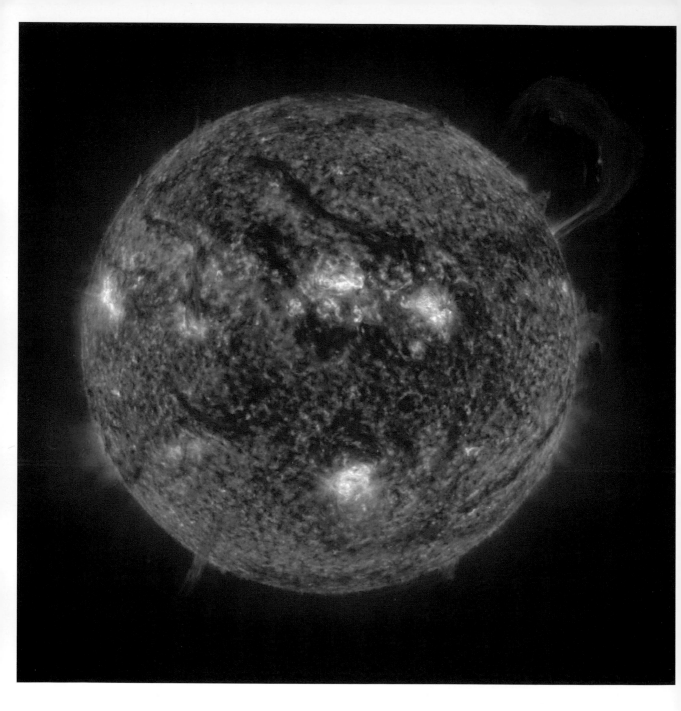

Extreme Ultraviolet Imaging
Telescope (EIT) image of a huge
solar prominence, taken from
the SOHO spacecraft.

Earthbound telescopes. Hundreds of thousands of plumes of gas, called spicules, constantly surge above the Sun's surface, leaping high into the solar atmosphere at hypersonic speed. Streamers of bright gas, prominences, reach still higher above the solar surface, forming spectacular arches large enough to encompass the giant planet Jupiter. Sometimes chunks of the Sun hurtle from the Sun into space, forming great tongues of solar debris that disrupt electronic communications and cause aurorae to fluoresce in our atmosphere. Scientifically, such huge discharges of solar material into space are known as coronal mass ejections.

In every sense, the Sun makes the solar system tick. On Earth, for example, it is responsible for nearly every variation of the weather and climate. The Sun radiates its energy in all directions. When we consider that the planets are mere specks of matter in a solar system dominated by the Sun, we can understand that only a tiny percentage of radiated solar energy reaches Earth. Yet, working in partnership with the natural rotation of our planet, and its tilted orbit, the Sun's radiation helps separate night from day, winter from summer, and the tropics from the temperate and polar zones. The uneven distribution of solar energy causes wind when hot air rises and cold air moves in to replace it; similarly, it helps move the great ocean currents. Its energy is used by the green plants on which we are dependent for oxygen, and food.

Even in a world dominated by science and technology, we feel a great wonder at the Sun. Life as we know it would be simply inconceivable without a star like ours.

The Beginning

We remember the great Renaissance astronomers as heroes of their day, iconoclasts who dared to say that the Earth circled the Sun, and that the telescope revealed more truth than the books of antiquity. But Galileo would never know of the condensation theory of galaxies, and was long buried when its first proponent was born in Germany. The newcomer was Immanuel Kant (1724–1804), not a true scientist but a philosopher who is credited with conceiving the most widely accepted (and incidentally the oldest) theory of present-day cosmology.

Kant was one of those incredible prodigies of history, a frail saddler's son with a deformed shoulder and a taste for theology, who was idolized by the young scholars of his time. He was intrigued by connections he saw between physics and metaphysics, writing that 'two things fill the mind with ever new

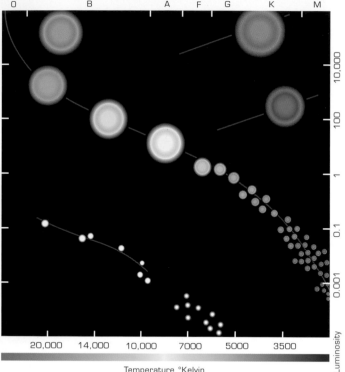

O B A F G K M

10,000
100
1
0.1
0.001

20,000   14,000   10,000   7000   5000   3500

Temperature °Kelvin

Luminosity

On the Hertzsprung-Russell diagram each star's position corresponds to its luminosity and its temperature. The vertical scale represents luminosity compared to our Sun's brightness, while the horizontal scale represents the star's surface temperature. 90 per cent of all known stars fall along the curved diagonal line called the main sequence. Above this line, at the top of the diagram, lie the massive and short-lived supergiants. Below the main sequence one finds white dwarfs – stars that have exhausted their nuclear fuel. Stars are not confined to the main sequence throughout their lives. For example, our Sun will move off the main sequence, above and to the right of its current position, as it exhausts its supplies of hydrogen and helium and swells to become a red giant. Finally, having burned all its fuel, it will sink below and to the left of its original position as it joins the line of white dwarfs.

and increasing admiration and awe, the oftener and more steadily they are reflected on: the starry heavens above me, and the moral law within me.' He must have been a strange man, who could write such words yet who, we are told, was unaffected by natural beauty, poetry, or music.

Kant, while not a mathematician, proposed that the primordial gas clouds not only coalesced into stars by falling in upon themselves, as posited by Isaac Newton in 1692, but also generated heat and rotation in the process. He visualized other 'island universes', similar to the Milky Way, existing within the cosmos.

Many years later, Kant's simple theory was backed up after Ejnar Hertzsprung of Denmark (1873–1967) and Henry Norris Russell of the United States (1877–1957) put together their cataloguing systems for the stars. Hertzsprung's and Russell's systems support the idea that star systems originate with the condensation of a scattered gaseous medium, its material provided perhaps by the supernova explosion of a giant star.

The Hertzsprung-Russell diagram, first published in 1913, is a chart that maps star temperature on one axis and star brightness on the other. The temperature is divided into 'spectral classes' labelled from left to right (hottest to coolest) with the perplexing sequence O-B-A-F-G-K-M. (Astrophysicists use the mnemonic 'Oh Be A Fine Girl, Kiss Me Right Now, Sweetheart (OBAFGKMRNS)' to remember this sequence, although the R, N, and S classes of stars have long since been abandoned.) Because a star's colour depends on its average surface temperature, stars also change colour from blue through white and yellow to red as one moves hot to cool. Our yellow Sun is classed as a G2 star (numbers 0 to 9 are used to indicate subdivisions within each spectral class), and is further defined as a dwarf, being of average brightness for its colour. In most aspects, the star we orbit is magnificently ordinary.

The traditional view is that, at its birth about 4.6 billion years ago, the Sun was surrounded by a churning cloud of matter, the solar nebula. As fragments of solid matter collided and stuck together, some eventually grew into larger

Computer rendering shows the evolution of a giant planet within a protostellar disc. As a Jupiter-mass planet forms, it excites density waves in the gas that push the gas away from the planet, forming a gap. High red bumps indicate high surface density, green is the original, unperturbed density, and blue is low surface density. Once the gap forms, there is no longer any gas left to accumulate on the protoplanet, and it stops growing.

aggregations called planetesimals, colliding with each other in turn and also pulling in large amounts of gas from the nebula around them. As the Sun became brighter, the solar wind – electrons and ions expelled from the corona and zapping along at up to 1,000 kilometres (620 miles) per second – blew away these thick, dense atmospheres and they became smaller, forming the terrestrial planets we know today.

Astronomers first realized the solar wind existed when they observed that comet tails do not stream behind them, but rather always point away from the Sun, blown in that direction by the constant stream of particles. The early planetesimals might well have looked like giant comets as their thick gas envelopes were stripped away. Further from the Sun, the solar wind would have been weaker, and less able to blow away the gas, resulting in the giant planets that orbit further out in the solar system.

Millions of small fragments from the planetesimal era remain today – for example the stony asteroids in the belt between Mars and Jupiter, and the icy comets that glide majestically across our skies, commemorating their finders with names like Halley, Kohoutek, Hale-Bopp, and Shoemaker-Levy.

How did all this begin? What existed before there were galaxies and solar systems? To form an understanding of the grand design of which the Sun and solar system are part, we must turn to cosmology – the study of the universe as a physical system, how it began and how it continues. For most of the twentieth century, cosmology was dominated by two rival hypotheses – the Big Bang theory and the Steady State theory.

Conceived by the Belgian astrophysicist Abbé Georges Edouard Lemaître (1894–1966) and developed by the American George Gamow (1904–1968), the Big Bang theory holds that the universe originated from the explosion of a single gigantic, super-dense 'singularity'. It is supported by the observation that the light from all distant galaxies exhibits a 'redshift'. Redshift is a Doppler effect, a change in frequency similar to the one we hear when listening to an emergency siren approaching or retreating. Light waves from a receding source reach us at a lower frequency than if the source was stationary, and so appear redder than expected. Conversely, the shift to higher frequencies caused by an approaching light source creates a 'blueshift'.

Doppler shifts in light can be measured accurately by looking at the very specific frequencies of light emitted and absorbed by certain atoms and molecules. Because galactic light sources appeared redder, we learned that the universe is rapidly expanding. If this expansion was traced backward, clearly everything must once have been much closer. Ultimately, billions of years ago, the entire Universe would have converged on one point – the singularity of the Big Bang.

Redshift also shows that the farther away an object is, the more rapidly it is retreating, and this became a handy shortcut to measuring cosmological distances – by measuring a galaxy's redshift, one can work back to its distance. However, in the late 1990s, scientists found an independent method of measuring the distance to these faraway galaxies, by searching them for a type of supernova that always shines at a certain absolute brightness. They found that the supernovae were fainter than expected, so the galaxies must be further away than their redshifts indicated. This means that the galaxies are moving outward at an increasing rate. Far from gradually slowing down due to gravitational attraction between galaxies, the universe's expansion is accelerating, indicating some mysterious anti-gravity effect, nicknamed 'dark energy', is at work in the cosmos.

The Steady State theory, now mostly of historical interest, holds that the universe is uniform in space and unchanging in time; that it has always expanded, with no beginning or end, always will expand, and will always maintain a constant density. Thomas Gold (1920–) of Cornell University and the Viennese-born British mathematician Hermann Bondi (1919–) first proposed the theory as an alternative interpretation of the universe's expansion in 1948. It was later refined by Fred Hoyle (1915–2001) and Dennis W. Sciama (1926–1999) of Cambridge University. (Sciama is also known for his statement, 'For my part, I believe that the aim of science should be to show that no feature of the Universe is accidental.')

Evidence gathered over the past few decades has clinched the case for the Big Bang theory, showing how, in the first three minutes of creation, it created large amounts of deuterium (heavy hydrogen with the same chemical properties but double the mass of the normal atom). Measurements made with the U.S. National Radio Astronomy Observatory 12-metre radio telescope revealed that deuterium is less abundant in the centre of the galaxy than the galaxy as a whole, suggesting that these heavy hydrogen atoms are being consumed by stars as they rain in from some external source. Stars do not produce deuterium, they consume it. So where does the stuff come from? As yet we have only one plausible answer: the Big Bang.

Another vestige of the Big Bang is ionized helium, which theory says should be found in interstellar space but not in stars or galaxies. Convincing evidence that this helium is indeed there emerged in 2001 after analysis of spectra obtained by the Far Ultraviolet Spectroscopic Explorer (FUSE) satellite.

The scale of the universe, already difficult to comprehend, has been pushed relentlessly outward as a result of improved telescopy. In 1995, Robert Williams, director of the Space Telescope Science Institute in Baltimore, told his astronomers to aim the Hubble Space Telescope, a joint project of the U.S. National Aeronautics and Space Administration (NASA) and the European Space Agency (ESA), into a section of sky that seemed almost starless from the ground. For ten days the telescope collected the tiny streams of photons that came from that region. The resulting image was stunning, referred to with a certain awe as the Hubble Deep Field. In a single frame it is a view of no fewer than three thousand galaxies, some of them older than any seen before. In 1998, NASA revealed Hubble images of ancient galaxies that were formed perhaps less than a billion years after the creation of the cosmos.

One of the main attractions of the Steady State theory was that it offered a universe without a beginning, and therefore avoided the awkward questions of what triggered the Big Bang and what might have come before it. In recent years, mathematics, string theory, and the discovery that ours is a flat universe have come up with a variety of possible answers. There could exist many universes – a multiverse in which the cosmos we inhabit could co-exist with others in many-dimensional space. New universes might emerge outward

You, Me, and the Sun

Philosophically speaking, Earth and her sister planets are mere knots of matter in the outer atmosphere of the Sun. Thus, after millions of years it is only natural that Earth's life forms have come to utilize solar energy in a steadily increasing number of ways. The Sun is part of us; we, in some infinitesimally small way, are part of the Sun.

Apparently, individual human beings are not very much affected by the solar cycle – protected by the atmosphere, the human body shows no reaction to the strongest solar storm. Some people have tried to link solar activity with trading patterns on the stock market and even with the admission of patients to mental hospitals. But neither theory seems to have any basis in fact.

Scientists have learned, however, that the average world temperature may be higher when solar activity is at its maximum than when it is at its minimum. This naturally has some effect on agriculture; in fact, the vintage years that produce particularly good Burgundy wine also happen to be years of high solar activity.

Though solar storms may come and go without our noticing them, normal sunlight can have pronounced effects on our bodies, some beneficial and some harmful. We know from personal experience that the Sun's radiation can be extremely powerful. It can cause severe sunburn by bombarding the body with extreme ultraviolet radiation, bring on sunstroke by overwhelming the body's natural cooling system, and, especially in people whose skins do not tan easily, promote a type of cancer – melanoma – by irradiating cells in the skin.

At the same time, it has been proven that our bodies can benefit greatly from moderate amounts of ultraviolet light; in the past, sunshine was prescribed as a cure for rickets, a childhood disease in which the bones are softened from lack of calcium. Experiments show that there is a photochemical reaction between the Sun's ultraviolet rays and cholesterol and other related substances in skin tissues. The reaction produces vitamin D3, and the vitamin helps supply the calcium needed for proper bone formation.

These of course are just small, subtle examples of the ways in which our species has adapted to our environment. On a much larger scale, it has been surmised that solar radiation, as well as cosmic radiation and that from radioactive material in Earth's crust, plays an important role in causing genetic change.

through black holes, those huge gravity sinks from which no light or matter can escape. Or repeated Big Bangs might occur far outside our three-dimensional model of the cosmos, as multi-dimensional regions of space called 'branes' crash together every trillion or so years and trigger violent expansion.

The Great Yellow Dwarf

Compared with our planet, the Sun is a giant – 1.4 million kilometres (864,000 miles) across as opposed to the Earth's meagre 12,800 kilometres (7,900 miles). Yet among other stars it is strictly average, a yellow dwarf rotating with

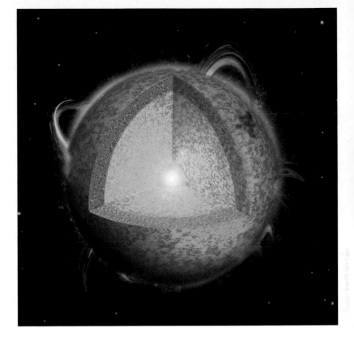

The Sun's massive core is topped by its brilliant photosphere and the relatively thin chromosphere. Looping prominences follow the curve of magnetic fields emerging from the star.

its captive planets around the centre of the Milky Way at the rate of one circuit every 250 million years. Many of our neighbouring stars in the Milky Way are far more impressive than the Sun. Betelgeuse in the constellation of Orion, for instance, is a red giant with a diameter of more than a billion kilometres (600 million miles), shining 9,000 times brighter than the Sun. Deneb, in the constellation Cygnus, is a somewhat smaller white supergiant, but shines with the light of a quarter of a million Suns.

Scientists gauge the mass of the Sun by a mathematical formula that decrees that a satellite body (such as Earth) can remain in stable orbit around another body (such as the Sun) at a known velocity and distance only if the central body has a given mass. Thus, if we know the diameter of Earth's orbit plus its mass and velocity, we can compute the mass of the Sun. At 2×10^{27} tonnes, it is 333,000 times greater than the mass of Earth. By comparing the mass of the Sun with its apparent volume, scientists have calculated that its average density is only 1.4 times that of water; for comparison, on average Earth is 5.5 times more dense than water.

The visible surface of the Sun is called the photosphere. It is characterized by granulations that have been compared to rice grains or cooked oatmeal. The granules, which are in fact the tops of convective cells in the act of bringing energy to the solar surface, measure up to 1,500 kilometres (950 miles) across. When the brilliant photosphere is blocked out by the Moon in an eclipse, the chromosphere shows as a flimsy pink rim around its edge; bright looping

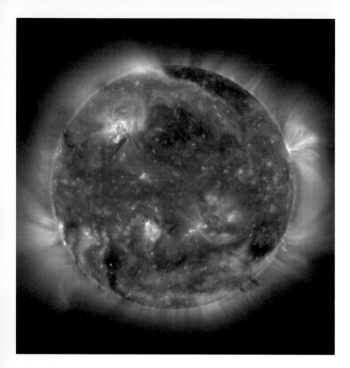

This composite image, made by combining EIT (Extreme Ultraviolet Imaging Telescope) images from the SOHO spacecraft, reveals solar features unique to each wavelength. The EIT images are produced in black and white, and later colour-coded for easy identification. For this image, nearly simultaneous images from May 1998 were each given a colour code (red, yellow and blue) and merged into one.

prominences hang over its surface; and the corona is a faint and diminishing halo that spreads far into space.

Optical filters, which allow the passage only of certain wavelengths (colours), reveal a great deal more detail. An unfiltered picture of the Sun shows the relatively featureless disc of the photosphere, marked with sunspots. In red hydrogen light we can see the sunspots of the upper chromospheric layers in great detail, with dark spicules radiating away along their magnetic pathways, and bright plages, from which solar storms (flares) can grow. An image of the Sun taken in plain yellow light, penetrates deeper into the photosphere, imparting greater contrast to its granular texture. Images taken from satellites, away from the Earth's atmospheric screen, can probe the Sun in other wavelengths, such as those of the far ultraviolet, where prominences show up in bright contrast against the background sky.

We can calculate the Sun's temperature, density, and pressure at other levels from information on mass, diameter, brightness, and chemical composition. These calculations indicate that the pressure produced by gases at the Sun's core must be around 300 billion times that exerted by the Earth's atmosphere at sea level. And to produce this pressure nearly half a million miles below the Sun's surface, the laws of physics decree that a gas must be heated to about 15 million °C (27 million °F). In his book *The Universe Around Us*, Sir James Jeans estimated that a pinhead of material at this temperature would radiate enough heat to kill a man one hundred miles away.

As we might expect, the Sun's internal temperature does decrease with distance from the centre, to about 5,800°C (10,500°F) at the photosphere. But there is an as-yet-unexplained reversal in the pattern above the photosphere. In the lower corona, the temperature rises to more than a million degrees.

The cause of this phenomenon remains elusive and haunts many a solar astronomer. One widely accepted theory holds that the mechanical energy generated in the seething mass of super-heated photospheric gases gradually works its way outward through the action of the Sun's magnetic field. Then it emerges into the solar atmosphere, where it is released as thermal radiation.

The Sun consists almost wholly of ionized hydrogen and helium. Even though these are the two lightest elements we know, the material at the Sun's core is more than ten times heavier than lead, with a density of 150 grams per cubic centimetre (4.25 tonnes per cubic foot). Such great densities are reached because the atoms only consist of their atomic nuclei, stripped by the high temperatures of their orbiting negatively charged electrons, which normally repel each other. If an atom were the size of a house, its nucleus would be a mere pinhead at its centre, so atomic nuclei can be packed much closer together than atoms in a normal gas while still remaining fluid. Because the Sun is fluid, it can spin faster at the equator (one revolution every 26.8 days) and slower at the poles (up to 35 days per revolution).

Today, after nearly a century and a half of solar spectroscopy, scientists know the Sun contains most of the elements present on Earth, though in very small amounts compared with the hydrogen and helium that dominate its composition. Hydrogen atoms are about 10 times more plentiful than helium and 1,000 times more abundant than oxygen, nitrogen, and carbon – as a result the Sun's mass is a little over 75 per cent hydrogen and a little under 25 per cent helium.

It must be remembered that the Sun is so intensely hot that these elements and all others exist in a state quite different from that which they assume on the cooler Earth. We may think of carbon as a block of charcoal, of magnesium as a lightweight white metal, and of silicon as the sparkle in a pile of sand. But in the Sun they are super-heated and turned to gas, and the gas atoms are converted to ions as their electrons are stripped away.

The Fusion Factory

The Sun radiates 383 trillion trillion watts – about the same energy as 100 billion tonnes of TNT exploding every second. What is the miraculous process that powers it, and drives every spark of movement in the rest of the solar system? The idea that nuclear fusion provides the Sun's energy was first proposed by the British astronomer Sir Arthur Eddington (1882–1944) in 1925, but it remained for Hans Bethe (1906–) to describe exactly how four hydrogen nuclei (protons) fuse together to form one helium atom, two positive electrons (positrons), and radiation. Each second at the Sun's centre, an incredible 700 million tonnes of hydrogen is converted into 695 million tonnes of helium by this process. Thus, every second, it converts the remaining five million tonnes directly into heat and light, according to Einstein's famous equation $E=mc^2$ (energy = mass x the speed of light squared).

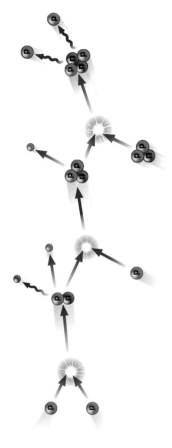

The Proton-Proton Cycle. The nuclear fusion process that fuels our Sun and other stars with core temperatures of less than 15 million °C (27 million °F).

The key to Bethe's theory is that heat and pressure deep inside the Sun are so great that protons (the single-particle, positively charged hydrogen nuclei) can hit each other so hard that they fuse together. In Bethe's proton-proton reaction, as two protons collide one spontaneously transforms into an uncharged neutron, getting rid of its excess positive charge by emitting a positron (the positive version of an electron), a gamma ray, and a particle called a neutrino. The surviving proton and neutron make up a deuterium (heavy hydrogen) nucleus, and when they collide with another proton they form a helium-3 nucleus (a light version of helium containing two protons and a neutron), releasing gamma ray energy at the same time. When two helium-3 nuclei collide, they form helium-4 (with two protons and two neutrons), and re-release two excess protons. For the net loss of four hydrogen nuclei, the Sun produces a helium nucleus and a considerable amount of energy.

The chain outlined above is the most common of several alternatives – others can result in the formation of light elements such as lithium and beryllium, while a completely different mechanism, which uses carbon, nitrogen, and oxygen nuclei as catalysts for the fusion process, takes place at the highest temperatures and dominates in stars more massive than the Sun.

The neutrinos produced in huge numbers by the fusion process are something of a mystery. They are tiny, possibly massless, neutral elementary particles that interact with matter via the weak nuclear force – a property that allows them to pass through the Sun and the Earth (and us) without being noticed. It's estimated that each of us receives 400 trillion neutrinos from the Sun every second, and billions more from natural and artificial radioactivity. Because our bodies contain the beta radioactive atom potassium-40, we even radiate neutrinos ourselves – about 340 million of them per day!

Quite naturally, astronomers wanted to detect neutrinos to confirm their ideas about nuclear fusion. Duly, they built detectors capable of finding them, but neutrinos come in different flavours, and those flavoured with the mark of the Sun turned out to be rarer than the calculations say they should be.

The first neutrino trap was a great 600-tonne vat of cleaning fluid located deep within a lead mine in South Dakota. Radiochemical analysis of the fluid found only a third of the solar neutrinos that theory said it should. A Japanese experiment that detected Cerenkov radiation (photons produced when particles impinge at a speed greater than the speed of light in water) brought the figure up to 50 per cent. Solid-state detectors eased it up a tad more.

Clearly something was needed to break the deadlock in this 'Solar Neutrino Problem'. Perhaps the flavours of some neutrinos changed after their escape

Visitors from Space

A vast variety of 'visitors' continuously invades the Earth from the Sun. This high-powered guest list is made up of electromagnetic radiation, which travels through space in waves at about 299,324 kilometres (186,000 miles) a second, along with protons, electrons, and other charged particles.

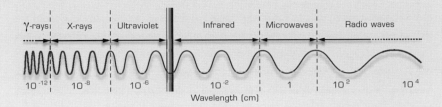

| γ-rays | X-rays | Ultraviolet | Infrared | Microwaves | Radio waves |

10^{-12} 10^{-8} 10^{-6} 10^{-2} 1 10^{2} 10^{4}

Wavelength (cm)

We can only sense a tiny part of the EMR spectrum – there is literally more to light than meets the eye. Although we ascribe names to various sections of the EMR spectrum, there are no actual hard 'boundaries' between each spectral region, just a continuum of smoothly changing wavelengths. Likewise there are no actual 'ends' to the EMR spectrum either – we just reach the practical limits of what we can measure.

The EMR spectrum is divided into an infinite number of wavelengths, ranging from gamma radiation to long radio waves. Between these two extremes in sequence from the super-short waves of gamma radiation are x-rays, ultraviolet rays, visible light, infrared, microwaves, short radio waves, medium radio waves, and long radio waves. Under the quantum theory, this radiation is said to be made up of elementary units called photons – a term commonly used to describe EMR from gamma rays through infrared.

If gamma radiation could trace a recognizable pattern across a sheet of paper, it would describe a series of waves so incredibly fine that several trillion peaks and troughs would appear in a single inch of travel. Radio waves, on the other hand, may be the length of several football fields.

All types of EMR are radiated into space by the Sun, and all reach the vicinity of Earth. But only visible light and low-energy ultraviolet penetrate the atmosphere with efficiency. Lower-energy radiation undergoes considerable filtering within the atmosphere, and higher-energy waves – which would be destructive to life – are absorbed before they reach Earth's surface.

Most of the charged particles from the Sun – which, unlike photons, have mass – are protons, positively charged particles found in atomic nuclei. They are usually paired with an electron to form a hydrogen atom, but the two are split apart within the Sun and ejected separately into space.

About 98 per cent of the charged particles cast off by the Sun are electrons, protons, and alpha particles (ionized helium atoms). Other charged particles produced include ionized atoms of titanium, calcium, strontium, magnesium, and other elements. Like the lighter particles, they are positively charged, because they lack part of their normal complement of electrons.

from the solar core, and assumed another flavour that could not be detected by the various custom-made detectors. There are in fact three known types of neutrino – electron neutrinos made by the Sun, and lower-energy tau and muon neutrinos produced by atomic decay and in nuclear reactors.

This idea – that electron neutrinos from the Sun might oscillate into new forms while en route to Earth – was intriguing enough for a new type of detector to be installed at the Sudbury Neutrino Observatory, built by the Canadian government with U.S. and U.K. participation. It required the building of a giant tank two kilometres (1.2 miles) below the Earth's surface in an Ontario nickel mine, and filling it with 1,000 tonnes of deuterium-based heavy water. Heavy water is special – when they zap through it, all three types of neutrino will produce flashes of light, bright enough to be picked up by the thousands of light detectors that surround the tank.

Before long, the Sudbury researchers were able to announce that yes, neutrinos do oscillate; they do transform; and apparently they also have mass. Science was on its way to solving a pesky problem about solar fusion, and developing a new tool to probe the internal workings of our star.

Bethe's nuclear fusion cycle produces a continuous supply of gamma-ray photons, 90 per cent of which are generated nearly half a million kilometres below the Sun's visible surface. These photons then begin a long journey through the Sun, bouncing back and forth between the atoms of the solar interior. This journey may take ten million years to complete because of the extremely erratic, random walk taken by these particles of light.

With each collision, the energy of the photon decreases slightly, and increases its wavelength. As time goes by, it may be converted to ultraviolet, to visible light, to infrared, and to microwaves and other radio frequencies. On Earth, we experience these ranges of electromagnetic radiation (EMR) as heat and light, as ultra-violet 'tanning' rays or as radio static – even though the original gamma rays would be impossible to detect without special equipment.

The Explorers

The port of Cayenne in 1679 was a forbidding place. Capital of the three-year-old colony of French Guyana, it was hot and humid and infested with tropical pests – hardly the place for a pleasant vacation.

But one afternoon that year a large sailing ship from France tied up and put ashore a most unusual passenger. This man was obviously a gentleman, and his luggage was full of the latest astronomical instruments.

The scientist was Jean Richer (1630–1696), an associate of Giovanni Domenico Cassini (1625–1712), director of the new observatory in Paris. He had government funding and a royal commission. But why had he come from France to this far corner of the world?

Richer was about to make the first reasonably accurate measurement of the distance between the Earth and Sun – the figure now known as the astronomical unit (A.U.). At a prearranged time, he noted the position of the planet Mars with respect to the Sun while Cassini, in Paris, did the same.

Then he and Cassini compared their diaries. The apparent difference noted in Mars's position – known as its parallax – was compared geometrically, and the Sun's distance was computed from the distance between the points of observation and the angles described between the observation points, Mars, and the Sun.

These computations indicated that the Sun was 140 million kilometres (87 million miles) away – 9 million kilometres off the figure accepted today but far more accurate than the few million kilometres proposed by Aristarchus, the Greek astronomer best known for proposing, in 275 BC, that the Earth circles the Sun.

Richer made another important discovery in Cayenne. Because his pendulum beat more slowly there, he deduced that gravity was weaker in Guyana – apparently because it was farther from the centre of the Earth. This information was then used by Isaac Newton (1642–1727) and Christiaan Huygens (1629–1695) to show that the Earth is an oblate (flattened) sphere.

In the following 200 years, many similar measurements of the Sun's distance from the Earth were made, using Venus and Mercury and even the asteroid Eros, as well as Mars, as reference points. Measurements became progressively closer to the modern figure, but it was not until 1887 that Britain's Sir David Gill (1843–1914) – after observing the Sun and Mars from Ascension Island in the South Atlantic – obtained a value nearly identical to that which we accept today. His figure: a shade over 93 million miles. The distance of the Sun today is accepted at 149,598,000 kilometres (92,750,000 miles); since the Earth's path is slightly elliptical, this distance increases and decreases as our planet travels along its orbit.

Richer was one of many pioneer scientists whose work was enabled and accelerated by the invention of the telescope. Joseph von Fraunhofer (1787–1826) mapped the solar spectrum in the early nineteenth century. The cyclical nature of sunspots observed by Galileo was discovered in 1843 by a German amateur astronomer, Heinrich Schwabe (1789–1875), and refined by

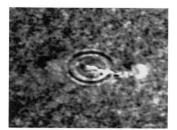

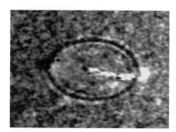

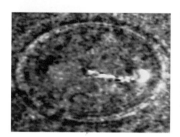

Solar flares produce massive seismic waves in the Sun's interior. Following a solar flare (centre to right in all frames), shock waves moving at 400,000 kph (250,000 mph) were tracked. After 70 minutes (final frame) they had formed a ring ten Earth diameters wide.

the Swiss Rudolf Wolf (1816–1893). In 1859, two Britons, Richard C. Carrington (1826–1875) and Richard Hodgson (1804–1874), were independently first to observe a solar flare. The fine structure of the solar atmosphere, studied during eclipses, began to be revealed in the final decades of the scientifically rich Victorian era. In 1891, the American George Ellery Hale (1968–1938) made it possible to photograph the Sun in any wavelength with his invention of the spectroheliograph. He later designed California's famed Mount Wilson telescope, and by observing the spectra of material around sunspots, showed that they have magnetic fields. The French astronomer Bernard Lyot (1897–1952) developed the coronagraph, which reveals the solar corona by creating an artificial eclipse, and a birefringent filter that made it possible even to make movies of solar activity.

One of the most important early instruments for observing solar phenomena was the spectrohelioscope, invented in the early 1890s by Hale and Frenchman Henri A. Deslandres (1853–1948). The device's name is derived from the Greek words *skopein*, to view, and *helios*, sun, plus the Latin *spectrum*, image. These scientists knew, as does anyone who has experimented with the optical properties of a prism, that white light is composed of all the colours of the rainbow, each with its own characteristic wavelength. The spectrohelioscope provided a way of viewing just one narrow band of wavelengths. It could be linked with a heliostat (Greek for 'sun-stopper') so that the Sun's image would be kept focused on the instrument for an entire day if desired.

Dark lines seen in the solar spectrum are signatures of specific elements (or molecules) present on the surface of the Sun. Detailed spectroscopic studies reveal the chemical content, temperature, pressure, and other atmospheric properties of the layers from which these lines originate.

Certain layers of the Sun's atmosphere are revealed by tuning the telescope's optical system to the right wavelength, or colour. This task, originally limited to use with a spectrohelioscope, can now be done by viewing the entire solar disc through specially designed filters. Certain wavelengths also indicate the presence of specific elements, such as calcium, and can be used to map temperatures in the solar atmosphere. Thus, one can make a composite map of three ultraviolet satellite images by assigning each temperature range a different computer-generated false colour – for example (photo, p. 20) red at 2 million °C (3.6 million °F), green at 1.5 million (2.7 million °F), and blue at 1 million °C (1.8 million °F).

A fairly new science, helioseismology, which maps the sound waves produced from deep inside the Sun, is beginning to give us exciting data that were never before available. Seismicity is triggered on Earth by earthquakes, but on the Sun it is a continuous rumble of acoustic waves, some of which peak on the order of once every five minutes as the star's mass expands and contracts. 'They occur ever so subtly,' says K. S. Balasubramaniam, an astronomer at the National Solar Observatory in New Mexico, 'like a ringing bell.'

Acoustic peaks can be detected by measuring changes in images of the Sun or by analysing the flux of energetic particles emanating from its surface. Temperature, composition, and motions deep inside the Sun influence the oscillation periods, so that at last we are able to map conditions in the solar interior from physical measurements and not from theory alone.

All these discoveries occurred, or could have occurred, before the first artificial satellite was lofted into space. They stand as tributes to a world scientific community that was dedicated to doing whatever was necessary to peer through the fog of atmospheric haze and learn all it could, even without the crystal clarity of space.

But alas, earth-bound solar observations still had their limitations. Because it is millions of times fainter than the photosphere, the corona can only be seen well in visible light from Earth when the glare of the photosphere is blocked by a natural or artificial eclipse. Furthermore, it is impossible to observe many important parts of the solar spectrum from Earth at all, because they are in the ultraviolet and X-ray wavelength ranges that are absorbed by the Earth's atmosphere.

Then at last the satellites swung into action. From 1962 to 1971, a series of Orbiting Solar Observatory satellites produced the best-ever photographs of the solar corona, imaged the Sun in the ultraviolet wavelengths that cannot be seen from Earth, and improved our understanding of the sunspot cycle.

More missions designed especially for solar astronomy followed on their heels – scores of them through the decades. Amazing photographs of the dynamic Sun have been created by Earthbound telescopes and more recently by the visible and ultraviolet instruments of SOHO (the Solar and Heliospheric Observatory), a joint project of the U.S. National Aeronautics and Space Administration (NASA) and the European Space Agency (ESA) launched in 1995. Koronos-F, equipped with radio, ultraviolet, and x-ray instrumentation, was launched on a similar mission by Russia and a group of international collaborators in 2001.

NASA has also been heavily involved in many other solar probes, such as the joint ESA/NASA Ulysses, the Cluster mission, TRACE (the Transition Region and

Coronal Explorer) and Yokoh. Ulysses, launched in 1990, flew over the Sun's poles for the first time, probing the heliosphere for information about the production of high-energy particles that threaten satellites and astronauts.

TRACE produced the first-ever maps of the magnetic fields that emerge through the visible surface of the Sun during its active periods, defining their structures and their related temperatures. Launched in April 1998, it allowed joint observations with SOHO.

The Yokoh ('Sunbeam') satellite, built by the U.S. and Japan with British participation, yielded more information on solar dynamics, charting solar flares through their X-ray and gamma-ray emissions. It was launched in August 1998 with a payload including spectrometers and two X-ray telescopes.

Cluster II was orbited with Soyuz launchers in July-August 2000, after the original spacecraft was destroyed in the disastrous 1996 explosion of an Ariane 5 system. It consists of four co-operating satellites that give an unprecedented 3-D view of the battle between the solar wind and the Earth's magnetic field, and its accompanying space storms. Administratively, it is remarkable in that it brought together more than 70 laboratories in the U.S., the European Union, Russia, China, Japan, India, Israel, Hungary, Poland, and the Czech Republic.

'Let no one underrate the ultimate versatility of these self-adapting machines the automata – their sensors and their artificial brains,' wrote the physicist and author William H. Corliss nearly four decades ago. 'Space probes are mechanical pathfinders for man's expansion into space. They scout the way and prove the equipment. As the stars themselves become attainable, unmanned automata will again spearhead the assault.'

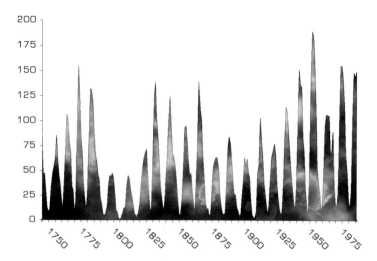

Daily observations of sunspots were started at the Zurich Observatory in 1749. By 1859, enough data had been collected for Heinrich Schwabe to discover that the number of sunspots waxes and wanes within approximately an 11-year cycle. At the 'solar maximum' there are more solar flares, more sunspots, and more solar quakes.

Sunspots, Prominences, and Flares

Though sunspots have been observed and studied since about 2000 BC, it was not until Galileo's time that the telescope brought accuracy to solar astronomy. Today, thanks to the system of sunspot observation originated by Rudolf Wolf at Zurich Observatory, scientists can trace the frequency of these events back to the year 1750, thus gaining clues to the nature of the solar system over a span of two and a half centuries. The chart opposite illustrates a kind of double cycle, revealing an outline that resembles three great wave peaks with smaller waves between. Each peak marks a year of high sunspot activity; the smaller ones occur every 11 years, while the larger ones crescendo to a new maximum about every 90 years. If the chart continues true to form, the peaks of the current 11-year waves will gradually build up until they reach a maximum in about the year 2050. Then the cycle will repeat itself.

Because of the Sun's cyclic nature, solar scientists have coined the terms 'solar maximum' and 'solar minimum' to denote the active and quiet parts of the 11-year cycle. A long period of very low sunspot activity, called the Maunder Minimum, occurred in the latter half of the seventeenth century, bringing with it such unusual effects as the freezing-over of the Thames and Seine rivers.

Sunspots can measure as much as 50,000 kilometres (31,000 miles) across, and are centres of great magnetic energy, always appearing in pairs of opposite polarity. They are convenient indicators of the Sun's overall level of activity, and, therefore, of the influence the Sun will exert upon the Earth. For example, radio signals from the Sun can be directly related to sunspots. In fact a graph showing sunspot activity for a given period looks almost identical to a graph that shows solar radio emission for the same period. Radio signals can tell us of the existence of an otherwise obscured sunspot, and the sunspot itself may flower into a prominence or a flare that will disturb Earth's region of space.

Both prominences and flares appear as violent displays of luminous solar debris and generally may be seen as clouds of ionized matter standing out from the Sun's limb or as characteristic markings on the disc itself. But there is a difference. Prominences linger much longer on the solar surface than flares. And prominences are seen as cascades of debris that may move downward along magnetic lines to the solar surface, sometimes in the form of arches that can span many tens of thousands of miles. Thus the material that forms a prominence is largely confined to the solar atmosphere. On the other hand, flares characteristically leap out from a single point of disturbance and form

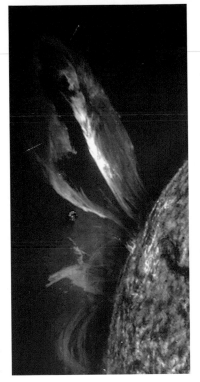

A large, eruptive prominence with an image of the Earth added for size comparison. This prominence from 24 July 1999 was particularly large and looping, extending over 35 Earths out from the Sun. Erupting prominences (when Earthward directed) can affect communications, navigation systems, even power grids, while also producing auroras visible in the night skies.

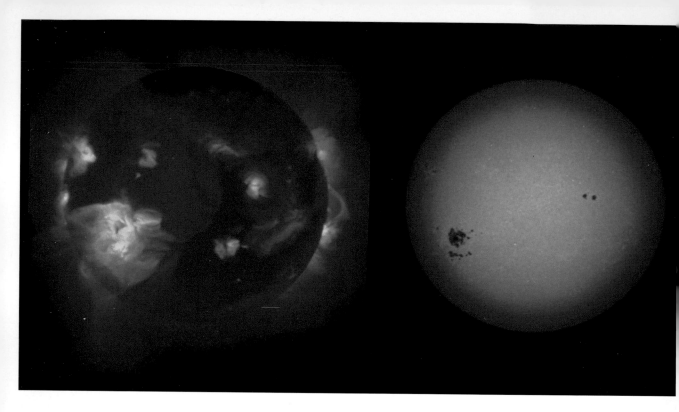

Comparison of false-colour X-ray (left) and visible light (right) images of the Sun reveals a correlation between sunspots and increased X-ray emissions.

the 'plasma tongue' that reaches far out toward the planets. The largest flares, called coronal mass ejections, pose a potential threat to space travellers, and may disrupt Earth communications.

Solar flares are classified in four categories, ranging from class 1, a small flare that may last for a few minutes, to class 4, a giant with a lifetime of up to an hour. Giant flares may cover a region the size of ten Earths and release energy equivalent to more than 10 trillion tons of TNT.

The first stage in the development of a solar flare is the appearance of a disturbed area called a plage in the chromosphere. Intense heating – possibly caused by magnetic field changes – occurs within the plage, and causes the production of microwaves (very short wavelength radio waves). The flare then begins to emerge from the centre of the plage. As it brightens to maximum intensity, giving off vast bursts of radiation, its internal heat approaches that of an exploding hydrogen bomb.

Next, a cloud of electrons flashes from the flare towards the edges of the solar atmosphere, carrying behind it a huge mass of superheated, charged gases torn from the solar surface. The gas and its spun-off electrons, shooting outward at supersonic speeds, then begin their trip through interplanetary space.

Hot gas frequently erupts from the Sun. One such eruption produced this glowing filament, which was captured by the Earth-orbiting TRACE satellite.

The filament, although small compared with the overall size of the Sun, measured over 100,000 kilometres (62,000 miles) in height,

If it is in the path of this plasma tongue, the Earth will soon experience a magnetic storm with its attendant radio interference and auroral displays. Electromagnetic radiation takes a scant eight minutes to reach Earth; high-energy protons race to us in as little as ten minutes, while their low-energy brethren will take up to forty hours to make the trip. The electron cloud will arrive at Earth in between twelve and ninety hours.

The Spectacle of the Aurorae

More than two millennia ago, the astronomers of China and Korea were keeping track of the northern lights, which periodically illuminate the skies with brilliant displays of green, blue, violet, and red spirals and curtains. And in describing a 'blood-coloured spectacle' that occurred in 349 BC, Pliny the Elder called the phenomenon 'a terrible portent, a conflagration falling Earthward'.

These natural fireworks displays, which have now been seen around the poles of several planets, may reach to heights of 200 kilometres (120 miles) or more on Earth. They were named in 1621 by the French astronomer Pierre Gassendi (1592–1655), who used the Latin words for 'northern dawn', *aurora borealis*. They also appear in the southern hemisphere, where they are called the southern lights or *aurora australis*.

The German-born scientist Johan Wilcke (1732–1796) determined two centuries ago that the aurorae were somehow linked with natural magnetism. It was also observed that aurorae often followed a flare-up of explosions on the Sun's surface. But it was not until the second half of the twentieth century that we had a satisfactory theory to explain what triggers the northern and southern lights.

Most of us have experimented in school science labs with a bar magnet and a bottle of iron filings. When the filings are scattered around the magnet, they arrange themselves along the magnetic field in a distinctive pattern, forming curving bands between the north and south poles. We now know that most planets, including Earth, have similar field lines forming what is known as a magnetosphere. Electrons and particles thrown off by the Sun are captured by these fields and drawn in spiral paths toward the poles.

We have all been attracted by the glow of neon lights – a glow provided by gas atoms whose electrons are raised to an 'excited' state and which return to their normal state by giving off the surplus energy in the form of light. A similar mechanism lights the aurorae. Particles strike the gas atoms and

Chasing the Moon's Shadow

When the Moon came between the Earth and the Sun and plunged our primitive ancestors into darkness, they reacted in fear, initiating ritual ceremonies to drive off the invisible monster that seemingly had swallowed the Lord of the Sky. But today, now that we understand and can predict solar eclipses, these occasions instead touch off periods of great scientific activity. A total eclipse can draw thousands of people from distant parts of the world.

The word 'eclipse' comes from the Greek *ekleipsis*, to leave out. When any source of radiation is blocked from view, it is said to be eclipsed; when we speak of a solar eclipse, however, we mean specifically that the Moon has come between the Sun and Earth. By an amazing coincidence, Sun and Moon appear almost exactly the same size in Earth's skies, so when the Moon passes in front of the Sun it hides the photosphere precisely. For those fascinated by odd words, this is a conjunctive *syzygy*, an alignment of three celestial bodies.

Because the match is so precise, a total eclipse can only be seen from a small strip of Earth's surface, along the 'path of totality'. These areas lie in the Moon's umbra, or shadow. The much wider area of Earth that is partially shadowed by the Moon and from which one may witness a partial eclipse is called the penumbra. A third type of eclipse, called annular, can happen when the Moon is at its most distant point from Earth during an eclipse – it cannot block out the photosphere totally, and instead leaves a thin ring of the Sun visible around its edges.

As an eclipse begins, the Moon begins to blot out the glowing solar disc, gradually moving across until all but the Sun's edge – the limb – is blocked from sight. At this point, ground observers can distinguish prominences surging from the Sun's surface. Then the Moon continues on its way as the shadow continues along the path of totality at 5,000 kilometres an hour (3,100 mph), eventually disappearing into night in some distant part of the globe.

Back in 1919, physicists took advantage of a solar eclipse to help prove Albert Einstein's theory of general relativity, which predicts that light bends as it passes through a gravitational field. An expedition led by British astronomer Arthur Stanley Eddington (1882–1944) journeyed to the island of Principe, West Africa, where they photographed stars lying close to the Sun during an eclipse. Comparison of these photographs with those taken at another time, and others made by an expedition to Brazil showed that the stars' apparent positions had shifted as their light passed through the Sun's intense gravitational field.

The *aurora australis* (or southern lights) photographed by the crew of the Space Shuttle *Discovery*.

molecules of the atmosphere, excite them, and cause them to radiate light of specific wavelengths. Different gases emit light of different colours: oxygen, for instance, characteristically produces green and red in the aurora; nitrogen produces blue-violet and deep red; and hydrogen also produces red. The intensity of radiation is dependent upon the density of the atmosphere, and there is a cut-off point at around 115 kilometres (70 miles) altitude, below which very few aurorae occur.

When the Sun Grows Cold

In the very far future, as the Sun becomes a giant star, the growth of its 'fusion zone' will envelop planets within the solar system, as the source of terrestrial life turns cannibal. First the Sun will envelop Mercury, then Venus, then very likely the Earth itself. For evidence of our fate, we need only turn to the telescope and to such giant stars as Mira and Betelgeuse, which already have diameters greater than the orbit of Mars.

Five or six billion years in the future, the Sun will have exhausted the available hydrogen at its core, and the fusion zone will begin to move outward. This will cause the outer layers to expand, and the drop in surface temperature will turn the Sun from yellow to red. Meanwhile, as the core slowly compresses under its own weight, it will eventually become hot enough for a new reaction to start up, fusing helium nuclei to make heavier elements. But helium fusion is short-lived, and eventually it too will have to move out from the core, triggering further expansion and turning the Sun into a red giant star, enveloping the inner planets. Earth's oceans will dry up, the atmosphere will drift away into space, and the remnants of our star will be surrounded by the ghost of all our seas and all its attendant comets and proto-comets – a colossal cloud of water vapour. As the fusion zone continues outward, the pressure and temperature will eventually drop until finally the reactions cut out. The outer layers will puff off in an enormous shell of gas – a so-called planetary nebula – while the dying Sun collapses under its own gravity to become a small, hot, incredibly dense white dwarf star.

The final stages of this drama are being enacted today throughout the Milky Way, among many thousands of so-called Mira variable stars, which can lose the equivalent of the Earth's mass of material every year. Astronomers can map this mass-loss process in a variety of ways – for example by detecting and measuring the radio signals given off by silicon monoxide (SiO) contained in the ejected gas.

Every once in a while a white dwarf in a binary system with another star will brighten dramatically in an enormous explosion – a nova or 'new star'. Novae happen where a white dwarf's gravity is pulling material away from its companion star. As hydrogen piles up on the white dwarf's surface, it eventually becomes hot and dense enough to trigger fusion, burning off in a short-lived but intense blast of radiation.

Massive stars have more impressive deaths – detonating in supernova explosions that can outshine an entire galaxy. These are rare in the Milky Way, but a frequent sight for telescopes observing the wider universe. In the current era, naked-eye supernovae were recorded in the years 185, 393, and 1006, 1054, and 1181. The remnant of the Taurus supernova of 1054 is today visible as the Crab Nebula, while in 1572 another of these stellar deaths excited the astronomical interests of Tycho Brahe. In 1604 Johannes Kepler, undergoing his transition from astrology to science, surprised his patrons by declining to invest that year's supernova with an astrological meaning.

As a star dies, it spawns the seeds of regeneration. Every time one of them falls apart, its fragments of gas and dust ride great shock waves into space, contributing to the universal store of thinly-spread matter from which new star systems like our own can take form.

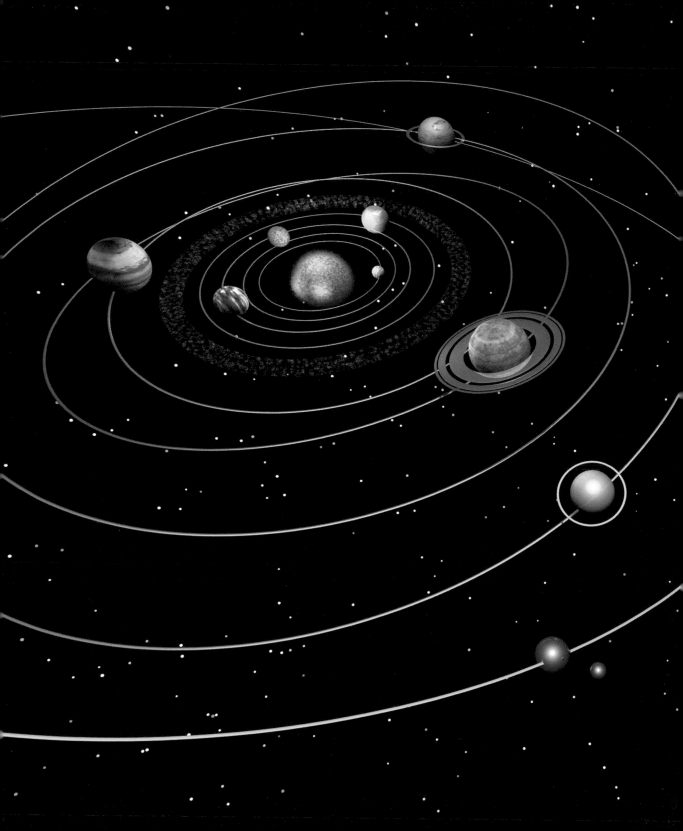

The View From Here

'I do not know what I may appear to the world; but to myself I seem to have been only like a boy playing on the seashore, and diverting myself in now and then finding a smoother pebble or a prettier shell than ordinary, whilst the great ocean of truth lay all undiscovered before me.'

Isaac Newton

We have been fascinated by other worlds since the moment the concept of planets – wanderers in the heavens – entered our minds thousands of years ago. Now we've been to the Moon; our robot explorers have mapped territories billions of miles from Planet Earth; and we're planning colonies on Mars. For our insights into worlds beyond our own we are indebted to those who first tried to understand the marvels of the solar system, who then developed the technologies that would enable a closer look at what's out there, and who today continue the work that permits us to follow our dreams into the sky.

Earth is one of four relatively small planets that plough like spaceships through the Sun's outer corona, leaving long wakes in the sea of solar radiation and particles that flows continuously out into the Milky Way. The other 'terrestrial' planets are Mercury, Venus, and Mars. Outward and away from us, in the spot where we might expect to find a planet, is that great zone of debris called the asteroid belt. Beyond this are the outer planets – Jupiter, Saturn, Uranus, Neptune, and Pluto – and the gigantic comet nurseries of the Kuiper Belt and Oort Cloud.

The terrestrial planets have three natural satellites: our own Moon, and the oddly shaped companions of Mars, Phobos and Deimos. By contrast, the outer planets have many moons. Jupiter has at least 39, Saturn 30, Uranus 21, Neptune 8, Pluto one – and life conceivably could exist on several of them.

The distance between Earth and the Sun is nothing compared with the enormous scale of the entire system. Mercury, Venus, Earth, and Mars are oddities in that they are relatively close together. The average distance between them is 37 million kilometres (23 million miles). Between the outer planets, the corresponding distance is about 1,260 million kilometres (800 million miles)! We could argue successfully that Mercury, Venus, Earth, and Mars are spheres of solid matter in the outermost layer of the Sun – and that humans are therefore not only residents of Earth, but of our star as well.

Even now, the planets retain the remains of the ancient Sun. Earth's own continents float upon a restless, spherical ocean of primeval solar materials that are kept hot by radioactive decay and a thick insulation of rock. From time to time these materials spurt forth and form volcanoes and other extrusions. Fires stoked at the beginning of time generate hot springs and geysers at the Earth's surface, and ooze their sulphurous life's blood at the oceans' great depths. There in the pitch darkness, they give life to creatures that are strange beyond belief.

Could we possibly have ignored the lure of these other worlds, the planets? In no way. Like the astronomers of history, we must follow our sense of wonder, for this is a central part of what makes us human. 'The challenge of the great spaces between the worlds is a stupendous one,' Arthur C. Clarke wrote half a century ago, 'but if we fail to meet it, the story of our race will be drawing to its close. Humanity will have turned its back upon the still untrodden heights and will be descending again the long slope that stretches, across a thousand million years of time, down to the shores of the primeval sea.'

Mapmakers: From Stone Tablets to Spacecraft

The urge to map the solar system – and the cosmos beyond – lies deep in the roots of humankind. We were mapping coastlines thousands of years ago, and it is likely that people were familiar with the skies earlier still. Our ancient ancestors kept time by observing the positions of the Sun and Moon, planted their crops in accordance with the Sun's position in the sky, synchronized religious ceremonies with celestial events, and navigated by the stars. We can see the remnants of our ancestors' stone observatories at sites around the world.

The sky, source of warmth and rain and wind, is also the place of portents. The second half of the eleventh century alone began with the explosion of the supernova that became the Crab Nebula, in 1054, then saw the apparition of Halley's Comet in 1066, and the formation of huge sunspots in 1077. For the skywatchers of New Mexico's Chaco Canyon this was a time of unprecedented celestial auguries, in which their skies were also visited by total solar eclipses in 1076 and 1097. At the beginning of this fifty-year period, their tiny city-state was blossoming; by century's end, it was in irreversible decline. Maybe astronomy – and superstition – had something to do with Chaco's downfall.

Simple early maps of Earth's features, such as those drawn on clay tablets by the Babylonians more than four thousand years ago, were based on the observations of travellers; these were perfected by the invention of latitude and longitude, by the use of instruments such as the compass and astrolabe, and by innovations that included the camera and aerial photography (see Chapter 5). England's William Gilbert (1540–1603) drew a map of the Moon in the last year of his life. But Galileo Galilei (1564–1642), with the aid of his telescope, was the real pioneer in mapping the solar system, discovering the moons circling Jupiter, the phases of Venus, and the sunspots, which he discussed in a fierce debate with a German contemporary who thought they were solar satellites.

Today the use of spacecraft for mapping the solar system – including Earth – is taken for granted. We've been at it for nearly half a century – the far side of the Moon was first photographed by the Soviet Luna 3 spacecraft in 1959. The Americans soon got in on the act, and flooded the scientific community with tens of thousands of photos of the entire lunar surface. In 1969, when astronauts first set foot on the Moon, they had detailed maps in their pockets.

By the turn of the century, at least part of every planet in the solar system had been mapped. The first detailed maps of Mercury were made in 1975; of Venus in 1978; and of Mars in 1972. The major moons of the gas giant planets were mapped for the first time in the 1970s and 1980s, and even surface details of Pluto have now been charted from the Hubble Space Telescope.

By some standards the terrestrial planets aren't much – they'd be negligible bits of rock were it not for the amazing accident that caused life to develop on at least one of them. Physically they're small potatoes compared with their four huge brethren on the other side of the asteroid belt. Those giants contain 99 per cent of the mass of the planetary system. They could be called the only 'real' planets – and Jupiter could be considered to be a failed, unignited sun. The outermost of the solar nine, Pluto, is minuscule by comparison, measuring only one fifth the diameter of Earth.

Many questions about the planets remain unanswered, and when we eventually find the answers, they are sure to generate many new questions. To pick just a handful:

Why, with a density much like the Earth, does Mercury resemble the Moon and contain little or no iron on its surface?

How can it be that, while Venus is covered in lava flows and other evidence of volcanic activity, there is no evidence of moving plates in its crust?

On our own planet, what is the connection between plate tectonics and global climate? How unique in the universe is life-friendly Earth?

Has life ever developed on Mars, and does it exist now? If so, where does the Red Planet's life survive, and what form has it taken?

Why did the Galileo probe encounter such dry conditions on Jupiter, and what has kept the storm that we call the Great Red Spot raging for centuries?

How did Saturn's rings originate, and why are they so much more dramatic than the rings around other planets? Why do the rings appear to have shadow-like 'spokes'?

Why is Uranus tilted almost 'on its side'? Why, unlike the other giant planets, does it radiate less heat than it receives from the Sun?

Why does Neptune have such furious winds when it has only a small amount of internally-generated heat? And why doesn't its magnetic field line up neatly with its poles, like those of Earth, Jupiter, and Saturn?

What is the actual size, shape, and density of little-known Pluto? How is it related to the icy worlds of the Kuiper Belt? And should it continue to be classed as a planet at all?

Early Space Scientists

Nicolaus Copernicus (1473–1543).

European scientists and philosophers who theorized that the planets move in orbits around the Sun were at first hounded by churchmen who preached Ptolemy's theory that everything revolves around Earth. Persian astronomers, safe from accusations of heresy, published early descriptions of a sun-centred system. Ironically it was a Polish cleric, Nicolaus Copernicus (1473–1543), who, without benefit of a telescope and apparently without access to the old texts, first showed the logic and mathematics of this model of the solar system, and in so doing led the way to a European renaissance in astronomy.

Copernicus was no self-promoter – he published no scientific works on his own initiative, and we are indebted to his disciple, George Joaquim Rheticus (1514–1574), for the fact that his theory reached the public at all. Copernicus was on his deathbed when the first copy of his master treatise, *De revolutionibus orbium coelestium* (*On the Revolutions of the Heavenly Spheres*) was brought to his side.

In Denmark, three years after Copernicus's death, a 14-year-old student was privileged, and astounded, to witness a total eclipse of the Sun. His name was Tycho Brahe (1546–1601), and the eclipse, predicted by Copernicus, inspired him to begin his brilliant career in astronomy. Tycho used simple instruments to show that the stars – of which he charted 777 – are not fixed in the heavens. He improved on Copernicus's predictions of planetary motions, but substituted his own model of the solar system, in which the planets orbited the Sun, but the Sun still went around the Earth. With royal patronage, Kepler lived an extravagant life and founded a great observatory, which he called Uraniborg after the muse of astronomy. He also lost his nose in a duel with a rival mathematician – another made of silver was substituted. Seven years after Tycho's death, the telescope was invented by a couple of spectacle-makers in Holland, and everything changed in a moment.

Tycho left his notebooks to his apprentice Johannes Kepler (1571–1630), a brilliant former divinity student who hitherto had lived more or less in the

Tycho Brahe (1546–1601).   Johannes Kepler (1571–1630).   Galileo Galilei (1564–1642).

great man's shadow. But Kepler turned out to be the better theoretician of the two. He was the first to dare suggest that planetary orbits might be ellipses rather than perfect circles – the crucial breakthrough that allowed accurate prediction of planetary motions and proved the Copernican theory right in spirit, if not in detail. Kepler devised three famous laws of planetary motion, showing mathematically that the planets travel around the Sun in elliptical

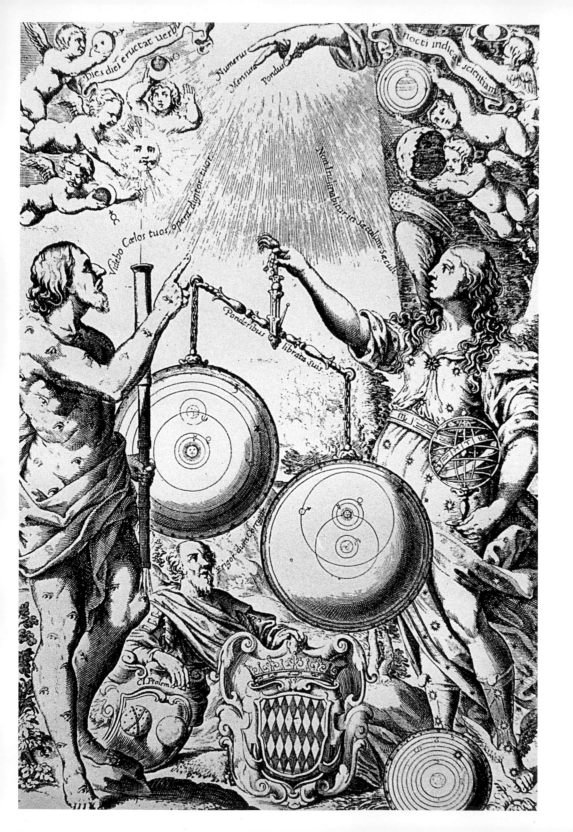

orbits with the Sun at one focus of each ellipse; that the speed of each orbiting planet varies with its distance from the Sun; and that the size of a planet's orbit is related to its orbital period by a simple equation.

Kepler's work is generally credited with laying the groundwork for Sir Isaac Newton, who in his *Principia Mathematica* of 1687 laid out brilliantly the relationship between gravitational forces and the mathematical rules of planetary motion. This was the genesis of the science that came to be called celestial mechanics.

The Italian contemporaries of Tycho and Kepler, meanwhile, were having a less pleasant time. Giordano Bruno (1548–1600) espoused Copernicus's idea that Earth circles the Sun and postulated that the universe is an infinite realm containing many worlds. He was excommunicated several times, finally to be burned at the stake. Justus Susterman's painting of Galileo, which hangs in the Uffizi Gallery, Florence, depicts a man in late middle age with a worried and watchful look on his face. For good reason. By that time this great pioneering astronomer – the first to use a telescope – had been twice condemned by the Church for heresy. Copernicus's work had been banned in 1616, and now Galileo was avoiding Bruno's fate by denying his true conviction that Earth circled the Sun. He kept quiet until his death, though there is an unconfirmed story that, when forced to declare his retraction, he still whispered 'Eppur si muove' ('And yet it moves!').

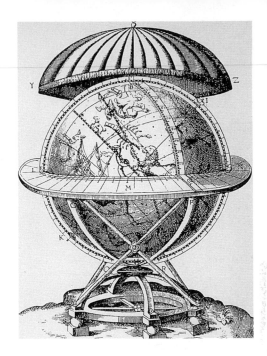

Above: The Great Brass Globe was one of the instruments at Tycho Brahe's observatory at Uraniborg. The globe, a perfect sphere 149 cm (58.5 inches) in diameter, was constructed from wood, brass and parchment and took 25 years to make. It showed the positions of the stars and constellations with great accuracy. This illustration is from Brahe's book *Astronomiae Instauratae Mechanica*, published in 1598.

Other Solar Systems

The Arecibo Observatory, a great spherical antenna carved into the limestone sinkholes of Puerto Rico, is the world's most sensitive radio telescope, able to gather the faintest signals from stars millions of light years away. Arecibo is also able to detect the flashes of radiation that shoot into space by electrons when they are captured in the magnetosphere of a dying star – a pulsar.

Since the first pulsar in a binary star system had been discovered at Arecibo in 1974, it was no coincidence that a young Polish-born physicist by the name of Alex Wolszczan found himself there sixteen years later, poring over the microwave traces from a newly-discovered, rapidly rotating radio source. It was 1,300 light years away in the constellation Virgo, and classified as millisecond pulsar PSR1257+12.

Left: The hand of God passes judgement on the relative merits of the Copernican (centre left) and Ptolemaic (lower right) cosmological systems; here, the Ptolemaic system has tipped the scales in its favour. Ptolemy described the heavenly bodies as orbiting around the Earth in complicated, circular epicycles. Copernicus suggested that the planets went around the Sun in simple orbits. Copernicus's solar system theory was suppressed by the Roman Catholic church until 1835. This engraving is from 1651.

Arecibo, the largest radio telescope in the world, is situated in a natural depression in the mountains of Puerto Rico. It can send and receive signals to examine planets and asteroids and also analyze Earth's upper atmosphere. To scan the sky, the dome is steered above the 305 metre dish. A recently installed new radar transmitter has increased Arecibo's power by 20 times.

There was something quite strange about the computer output. This pulsar did not send out an evenly spaced stream of pulses. Instead, the pulses undulated, apparently speeding up and slowing down. Wolszczan scratched his head and tried to figure out a likely reason. And he scratched his head again – there was no likely reason. The only thing that could cause this peculiar behaviour would be the existence of three planets orbiting the pulsar. If they existed, these planets, he calculated, would be roughly the same distance from the pulsar as Mercury is from our own Sun. The smallest would have a mass of just 0.015 Earths, and orbit its star in just 25 days. The others would be larger, with 3.4 and 2.8 Earth masses, in circular orbits of 66.6 and 98.2 days.

Not surprisingly, Wolszczan's theory was met with healthy scientific scepticism. But he checked and double-checked his findings, and had them cross-checked by a colleague at the Very Large Array telescopes in New Mexico. About a year after his discovery, he published the results in *Nature*. Two years later, after more painstaking analysis, he announced in *Science* the 'final proof that the first extrasolar planetary system has been unambiguously identified.'

By that time, the optical telescope specialists were hot on the trail. In 1994, C. Robert O'Dell of Rice University in Texas found that no fewer than fifty-six of one hundred young stars in the Orion Nebula were surrounded by discs of dust – flat silhouettes standing out against the bright cloud of material that surrounds each parent sun. Images of planets forming within those nebular discs, however, were hard to come by.

Observers look for traces of extrasolar planets (sometimes called exoplanets) by seeing if a particular sun wobbles on its course, as if tugged by the gravity of a planet. Alternatively, the star's brightness might decrease as a planet passes in front of it. They can also try to block out the light of the Sun and look for signs of planets in the light surrounding it, or search for dust rings that could be 'shepherded' in place by planets.

Then it happened. In October 1995, Michel Mayor and Didier Queloz of the Geneva Observatory announced that they had discovered a large planet orbiting 51 Pegasi, a star similar to our own Sun, 45 light years away. Mayor and Queloz had studied Doppler shifts in the star's light over a twelve-month period, using a spectrograph at the Haute-Provence Observatory in France. When unscrambled, their data revealed the planet's size (roughly that of Jupiter), temperature, length of year, orbital eccentricity, and proximity to 51 Peg – a puzzlingly close 0.05 A.U.

The Swiss team's discovery was confirmed later that month by independent U.S. teams. The first extrasolar planet had been discovered by an optical telescope, though no image was produced. All the data came from spectrographic variations in the light received from 51 Peg.

When the tenth extrasolar planet was announced, we suddenly knew of more planets outside our solar system than within it. Today, over one hundred of them have been discovered, one of which has the equivalent mass of 17.1 Jupiters! Such discoveries cause the experts to believe that solar systems may form quite differently around stars other than our own. The orbits of their planets are different in that they are very elongated; the large planets appear to orbit close to their stars; and (like our own Sun) their stars contain more iron than stars that appear to have no planets.

A system more like our own orbiting a Sun-like star, 55 Cancri, was announced in 2002 by a team headed by Geoff Marcy (University of California, Berkeley) and Paul Butler (Carnegie Institution of Washington). It features a Jupiter-like planet and theoretically could include a terrestrial planet as well.

Telescopes on the Kepler satellite, scheduled for launch in 2006, will search for the slight dimming of starlight as planets transit across the bright discs of their suns. Changes in brightness will indicate planet size; the period of the dimming will allow scientists to calculate the orbit size and estimate the planet's temperature. By then, the Space Infrared Telescope Facility, the final mission of NASA's 'Great Observatories' programme, will also be in orbit, hunting down planet forming regions in dust discs around nearby stars, as well as studying the early universe.

Tools for Astronomy

The Sun and planets have teased humankind's sense of wonder since the dawn of intelligence, but not much was known about them until Galileo made the first scientific use of telescopes.

'That which will excite the greatest astonishment by far, and which indeed especially moved me to call the attention of all astronomers and philosophers, is this,' wrote Galileo in 1610, 'namely, that I have discovered four planets, neither known nor observed by any one of the astronomers before my time.' He was referring to the four large moons of Jupiter – Io, Europa, Ganymede, and Callisto. He found his telescopic discoveries so wondrous that he exclaimed, 'Oh, when will there be an end put to the new observations and discoveries of this remarkable instrument!' No answer is yet in sight.

With the completion of Newton's *Principia Mathematica*, the stage had been set for the development of astronomy – a Greek expression that means 'the way of the stars', and currently defined in *Webster's New Collegiate Dictionary* as 'the science of the celestial bodies and of their magnitudes, motions, and constitution'. Telescopes (the word comes from another Greek term meaning 'far-seeing') collect photons of electromagnetic radiation from distant objects, and convert them into images. The same photons may be used to determine a variety of physical characteristics, from the speed at which a distant planet is moving towards or away from us, to a breakdown of the gases present in that planet's atmosphere. These photons may express themselves as visible light, such as one can see through a conventional telescope, or they may be of other electromagnetic frequencies, from gamma rays to radio waves.

Thus on Earth we can use optical telescopes, radio telescopes (which pick up radio signals from distant objects) and radar telescopes (the active version of radio telescopes, which bounce radio signals off planets and analyze the returning signals). Meanwhile, in nearby space, a variety of orbiting telescopes collect the short-wavelength radiations that cannot penetrate our planet's atmosphere. Telescopes remain the primary tools of astronomy, but their findings are now supplemented by a wide array of instruments that are sensitive to other physical phenomena, such as magnetic fields, temperatures, and pressure.

Galileo's use of the early optical telescope to study the heavens, the laws of Kepler, the gravitational theories of Sir Isaac Newton, and the ideas of thinkers like Immanuel Kant, all set the stage for closer physical examination of the planetary system. Then came the revolution of spectroscopy, through which Joseph von Fraunhofer (1787–1826) demonstrated that the chemical components of celestial bodies could be determined by examining the rainbow of colours separated by a prism from their emitted or reflected light. Pieter Zeeman (1865–1943) showed in 1896 that the same spectra could also reveal the strength of the magnetic fields present in objects.

Optical telescopes provided increasingly better images of the planets and stars as the twentieth century progressed, but were still limited by the photographic plates that were used to capture images and spectra at the end of the process. With the invention of the charge-coupled device (CCD) – today widely used in home video cameras – the sensitivity and versatility of detecting devices improved immensely, to a point where individual photons from the faintest light sources can be measured. The Hubble Space Telescope's main camera, for example, has a detector consisting of four postage stamp-sized CCDs, each with 640,000 individual 'pixels' that convert the light focused on them into an electric charge. The result is an imaging device that is a hundred times more sensitive than photographic film. Within the atmosphere, distortions caused as light waves pass through turbulence can also be erased by another relatively recent scientific marvel, adaptive optics. This technique senses unevenness in an incoming light wave and 'flattens' the wave with a flexible mirror that distorts it into the correct shape.

Modern imaging technology can enhance astronomical techniques that have been in use for years, including the observation of occultations – the temporary disappearance of a light source as another object passes in front of it from our viewpoint.

Occultations can provide valuable information – if you have your astronomical instruments set on a star as a planet moves into the field of view, you can gauge the diameter of that planet by measuring the time it obscures the star's image. If the light from the star winks or dims briefly, before or after it encounters the planet's disc, then perhaps it has been occulted by something else – a ring or a moon. Naturally, the closer the observer, the more precise the measurement. Detailed optical and radio observations make it possible to

measure not only the size of planets and their moons, but also to determine the extent and density of their atmospheres.

The characteristics of supernovae, quasars, pulsars, and other mysteries of the deep cosmos are mapped in detail by radio telescopes. These great listening posts trace back to 1933, when a young Bell Laboratories scientist, Karl Jansky (1905–1950), discovered radio static coming from the Milky Way. They developed through the work of others including James Stanley Hey (1909–), who discovered signals from our own Sun while he was working on early radar in Britain during World War II, and Sir Bernard Lovell (1913–) who built the world's first radio telescope at Manchester, England.

In the flowering of radar astronomy, radio beams from Earth were bounced from planetary surfaces to determine their roughness, size, and distance. The periods of rotation of Venus and Mercury could be learned from these radio echoes, and their basic features mapped. More recently radar has even peered through the clouds of Saturn's mysterious moon Titan.

The extreme sensitivity of radio telescopes is difficult to comprehend. At the time that he presented his *Cosmos* television series in 1980, the astronomer, writer, and science advocate Carl Sagan (1934–1996) estimated that 'the total energy received by all the radio telescopes in the world is less than a single snowflake touching the ground'. The metaphor is likely to remain valid for some time to come. The signals from distant objects are so faint that they require a special unit of measurement, the Jansky, defined as 10^{-26} watts per square metre per Hertz.

These revelations naturally brought up more questions than they answered – and the crispness of images from Earth-based instruments, even those on top of mountains such as the 4,205-metre (13,796-ft) Mauna Kea, Hawaii, was still limited. (The summit of this extinct volcano is graced with eleven giant telescopes, located above 40 per cent of the atmosphere, and renowned for its clear, pollution-free skies.)

Only spacecraft and orbiting telescopes could bring the ultra-high definition and close-up detail that astronomers demanded of their images of

Above: An occultation of Venus and the Moon over Mauna Kea, Hawaii. The 4,205-m (13,796-ft) high summit of Mauna Kea hosts the world's largest astronomical observatory, with eleven telescopes. Nine of them are for optical and infrared astronomy, and two of them are for submillimetre wavelength astronomy.

Below: Mars from an Earthbound telescope.

the solar system. When this dream began to become reality through the 1970s and 1980s, the results were truly marvellous. As the Jet Propulsion Laboratory's William I. McLaughlin wrote in 1989, nearly a decade before the Galileo spacecraft reached Jupiter, 'Our experience as Voyagers 1 and 2 reaped their harvest at Jupiter and Saturn beggars the powers of description and is even now difficult to comprehend in its extravagance.'

One of the most exciting moments in the history of optical telescopy occurred in April 1990, when the crew of Space Shuttle *Discovery* eased the Hubble Space Telescope (HST) into orbit 600 kilometres (373 miles) above the Earth's surface. A co-operative project of NASA and ESA, Hubble was designed to be maintained by Shuttle astronauts, whose first major task was to correct the system's blurry vision with new equipment in December 1993.

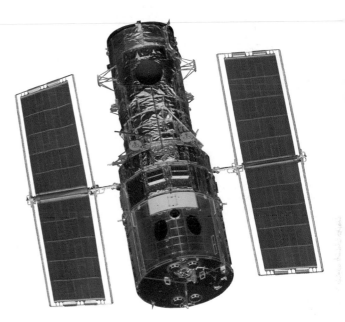

Above: The Hubble Space Telescope, famous for its photographs of the galaxy, is also searching for a new class of solar system object 5,000 times less massive than the Pluto and Charon twins. Dubbed 'mini-mes', they travel in pairs in the Kuiper Belt, a junkyard of icy bodies left over from the solar system's formation.

Hubble's array of cameras and instrumentation can be changed periodically. As of 2002, its instrument bays hosted the Wide Field and Planetary Camera-2 (nicknamed Wiffpick); the Space Telescope Imaging Spectrograph; the Near-Infrared Camera and Multi-object Spectrometer (NICMOS), with its super-cooled infrared detector arrays; and the state-of-the-art Advanced Camera for Surveys, capable of imaging a broad spectrum of light between ultraviolet and infrared.

The HST will come to the end of its design lifetime in 2009, to be replaced by a NASA-ESA instrument, the Next Generation Space Telescope (NGST). This optical and infrared telescope, fitted with an 8-m (26-ft) primary mirror, will image the universe as it looked when it was between one million and a few billion years old.

Below: Mars from the Hubble Space Telescope.

Many of the incredibly beautiful images of distant galaxies and nebulae, and the newer photos of a mottled Pluto, are the result of Hubble's remarkable optical technology. But these images don't just appear from the telescope ready for publication – the data streams from the HST must be run through a gamut of computer analysis to produce the spectacular end results. Some of the final results are even combined images from Hubble, spacecraft, and terrestrial telescopes. Take as an example the time-lapse colour movie of Neptune,

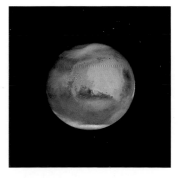

showing its huge jet stream, clouds and equatorial storms, put together in 1996 by scientists of the University of Wisconsin at Madison. These were not simply Hubble images or photos from the Infrared Telescope Facility (IRTF) on Mauna Kea. They came from both sources, in the form of images overlaid by computer. In 1998, Jet Propulsion Laboratory (JPL) scientists imaged a pair of huge Jovian ammonia-cloud thunderheads as they collided and joined together, using data merged from Hubble, the IRTF, and the Galileo orbiter. The movie was, literally, the best of three worlds.

But only spacecraft can give us a close-up view of the solar system's planets and their moons. Robot explorers bearing names such as Pioneer, Voyager, Galileo, Magellan, and Cassini offer us the nearest experience possible to actually being there ourselves.

By the turn of the century, astronomy was deemed to be the most prosperous branch of science around. Many new optical telescopes were in use and on the drawing boards (some with the aim of finding Earth-like planets in other solar systems). In December 2000, hundreds of people trekked to a remote Chilean mountaintop to dedicate two new telescopes, each built by the Carnegie Institution and equipped with main mirrors that measure 6.5 metres (21 ft) across. These huge telescopes are equipped to search out and analyze stars in the far reaches of our galaxy, the Milky Way. Yet, even as these celebrations were taking place, the University of California and California Institute of Technology were planning a telescope with a 30-metre (100-ft) mirror, dubbed the California Extremely Large Telescope. Meanwhile, the European Southern Observatory – which already operates the Very Large Telescope (a computer-linked collection of four 8-metre [26 ft] telescopes) – is planning an Overwhelmingly Large Telescope (OWL), which will have a 100-metre (325-ft) primary mirror. The name, at least, will be hard to beat.

The Lure of Being There

Just a few decades ago, remarkable discoveries were made with relatively inexpensive optical equipment, from deep within Earth's atmosphere. We charted the solar system in detail, even to the basic chemistry of Neptune's clouds. We found galaxies billions of light years away; fattened our dictionaries with new words like supernova and pulsar; and produced scholarly theories on the origin of the Universe. Telescopes are marvellous inventions that have brought us vast treasures of new information. Nothing will replace them. Yet it is in our nature to seek a closer view – and if possible to be there ourselves.

The Hubble Space Telescope was designed for in-orbit servicing and upgrading. Five days of space walks on Shuttle mission STS-061 in 1993 saw gyroscopes, magnetometers, solar panels and the Wide Field and Planetary Camera replaced. Most importantly, COSTAR (Corrective Optics Space Telescope Axial Replacement) was installed to correct the HST's spherical aberration of its main mirror, which had meant it couldn't focus properly.

As early as 1965, a working group of the U.S. National Academy of Sciences concluded that eventually the human body's reactions to long periods in space would be well enough known to permit travel to the inner planets. But this knowledge would not come easily, and it was clear from the start that the planets would be explored first with robotic systems.

Robot spacecraft must travel a variety of routes. Some are artificial Earth satellites, lingering in orbit like the Hubble Space Telescope and the craft that map Earth's weather patterns. Some orbit the Sun to support studies of solar phenomena; others perform 'flyby' or 'rendezvous' missions that seek new information about asteroids or comets. Others – described in detail in this book – travel along a solar orbit until they intersect another planet's gravitational field and are captured by it. They may then move on to another planetary destination, or they may release instrumented packages that land upon or penetrate the planetary surface. Some even escape the Sun's gravity altogether.

Humans have been mapping Earth and the other inner, or terrestrial, planets with spacecraft since the 1960s. By 1977, when the brilliantly successful twin Voyager spacecraft began their journey to the outer planets, a number of United States and Russian probes had already studied Mercury, Venus, and Mars.

Pioneer and Voyager took us out beyond the asteroid belt. Then, at the end of the twentieth century, a trio of remarkable spacecraft brought the outer planets much closer to our consciousness than had been achieved with their predecessors. First came the Hubble Space Telescope (1990 onward); after Hubble, the Galileo flight to Jupiter (1989–2003); and after Galileo, the Cassini mission to Saturn (1997–2008).

The Voyagers and their successor missions covered billions of miles while using only a tiny amount of fuel. In the space of a few years, guided through a near-miraculous game of celestial billiards by their human controllers, they greatly increased our knowledge of the solar system. Once upon a time the aurorae of Jupiter, the volcanoes of Io and the wrenching thunderstorms of Saturn were the stuff of science fiction. Now we have seen them for real.

The National Aeronautics and Space Administration (NASA), the European Space Agency (ESA), and their partners in space are actively exploring the solar system with the philosophy that planetary reconnaissance (carried out by flybys, orbiters, and atmospheric probes) will be followed by missions that return samples of planetary material to Earth. The thrust so far has been toward the most habitable worlds, as we look forward to establishment of bases on our Moon and (in particular) Mars.

NASA

The United States' National Aeronautics and Space Administration (NASA) officially began operations on October 1, 1958, after inheriting the earlier National Advisory Committee for Aeronautics (NACA), and other government organizations. NASA, in its own words, was formed 'as a result of the Sputnik crisis of confidence', following the successful Soviet satellite launch of 1957, and the failure of early U.S. attempts to reach space.

NASA had a workforce of about 18,000 at the time of writing, working on 'strategic enterprises' that included aerospace technology; human exploration and development of space; Earth science; space science; and biological and physical science. It is by far the world's largest space agency, though others, notably Europe and Russia, have taken an increasing share of the spacecraft and space launch business.

NASA currently runs dozens of space science missions, many of which involve international partners or other U.S. agencies. These run the gamut from the space shuttle and mammoth International Space Station to small Earth-orbiting sensor platforms and robot explorers across the solar system.

Headquartered in Washington, NASA has twelve centres, grouped into four categories – spaceflight, aerospace, space science, and Earth science. Mission Control is located at Johnson Space Flight Center in Houston, Texas, which takes over spacecraft operations immediately after launch with the assistance of ground stations around the world. Science experiments are run from the Jet Propulsion Laboratory (JPL) in Pasadena, California.

While each has its own specialities, the centres work together on strategic enterprises. There are four spaceflight centres – Johnson Space Flight Center (Texas), Kennedy Space Center (Florida), Marshall Space Flight Center (Alabama) and Stennis Space Center (Mississippi). Likewise there are four aerospace centres – Ames Research Center (California), Dryden Flight Research Center (California), Langley Research Center (Virginia) and Glenn Research Center (Ohio). JPL (California) and Goddard Space Flight Center (New York) specialize in space sciences and Earth sciences, respectively. There are also two test laboratories – White Sands Test Facility (New Mexico) and Wallops Flight Facility (Virginia). Although NASA contributes to some Earth-based astronomy, this is generally funded in the U.S. by the National Science Foundation and private foundations.

The European Space Agency

The European Space Agency started operations in 1975, two years after a meeting of ministers from ten European countries decided to combine the aims of the former European Launcher Development Organisation (ELDO) and the European Space Research Organisation (ESRO).

By the beginning of the new century, ESA had 15 member states in Europe, plus Canada as a co-operating state. Europe's space industries were employing 40,000 people directly and 250,000 indirectly, and were developing rapidly on multinational lines. As in the U.S., European space interests are consolidating rapidly, one example being the formation of Astrium from Matra Marconi (an alliance of French and British interests) and Dornier of Germany.

Unlike NASA, ESA did not inherit an infrastructure of government laboratories specialized in aircraft research – all its facilities were developed specifically for space research. When ESA began operations, member states naturally felt a strong desire to catch up with the U.S. and Russians, on whom they had to rely for launching their relatively small satellite programme (early attempts to build an all-European launcher had failed). Today, the agency has a full set of capabilities, including the Ariane 4 and 5 launchers, which have captured much of the commercial satellite launch market from the more established space powers. It enjoys an outstanding international reputation for the quality of its scientific work and has launched dozens of satellites and spacecraft to explore the solar system.

ESA's establishments and ground stations are scattered across Europe, with headquarters in Paris, France. ESTEC (European Space Research and Technology Centre) is at Noordwijk, in the Netherlands; ESOC (European Space Operations Centre) is in Darmstadt, Germany; ESA's main centre for Earth observation, ESRIN, is at Frascati, Italy; and EAC (the European Astronaut Centre) is at Cologne, Germany.

The Guiana Space Centre, CSG (Centre Spatial Guyanais), is located at Kourou, French Guiana. Following launch, ground stations swing into action at Salmijärvi (near Kiruna) in Sweden, Redu in Belgium, Villafranca del Castillo in Spain and Kourou. The agency also may use a tracking station in Perth, Australia, as well as other stations world-wide, as the missions require.

The Other Side of the Story

For the past few decades we have become accustomed to the idea that we are all citizens of 'Spaceship Earth'. Thanks to astronomy and the space programme, more people understand what it is to be children of the Sun, grandchildren of supernovae. Some hope this broadening perspective will help reduce tensions on our home planet. 'The crossing of space,' predicted Arthur C. Clarke, 'will turn man's mind outward, away from his tribal conflicts, and only thus will the world break free from the ancient cycle of war and peace.'

But it's expensive. The great German rocketry pioneer, Wernher von Braun (1912–1977), had a more practical perspective, which he expressed with wry humour. 'There is just one thing I can promise you about the outer-space programme,' he said. 'Your tax dollar will go farther.'

Ironically, the machines that might provide Clarke's freedom from tribal conflict came to maturity as a result of superpower competition. Until the end of World War II the balloon was the highest-flying 'space vehicle' in the scientist's stable. It was not until 1945 that the United States fired its first large scientific research rocket. Then, in 1957, Russia launched the first artificial Earth satellite, Sputnik 1, using rocket technology from its ballistic missile program. Sputnik was a small metal ball that was little more than an orbiting radio beacon, but these were the days of the Cold War between the Soviet Union and the West. The launch had a huge propaganda effect, launching humankind into the age of the satellite and the U.S. into a spirited competition for space supremacy with the Russians.

Suddenly America perceived that it needed spacecraft and a space programme, and Congress found ample money to fund it. Scientists on both sides were soon designing Earth satellites to photograph secret military installations. But this was not simply a military contest – it was also a scientific one, in which the U.S. soon gained the lead. In 1960, Pioneer 5 became the first interplanetary probe, measuring magnetic fields and particles far from Earth. In 1962, Mariner 2 became the first successful spacecraft to fly by a planet, Venus, and it was followed in 1965 with Mariner 4's fly by of Mars. Mariner 10 flew past Mercury three times in 1974 and 1975. Eventually, the Moon, Mars, and Venus were landed upon by robotic U.S. and/or Russian spacecraft. Meanwhile American Apollo astronauts visited the Moon over several years from 1969. The political importance of this programme was underscored when an audiotape of a 1962 meeting between President John F. Kennedy and NASA

Administrator James E. Webb was released by the JFK Library in Boston. 'I'm not that interested in space,' the President said. Then: 'Everything we do ought to be tied into getting to the Moon ahead of the Russians.'

Beyond the Asteroid Belt

Partly because of its closeness to Earth, partly because in some respects its structure and environment is closest to Earth's, Mars is the most-explored planet. More than 30 Mars missions have been attempted, most of which have failed, and a great many more are planned for the future (see Chapter 6). But what of the giant planets on the other side of the asteroid belt?

Back in the late 1960s, guesses as to what we might find in that mysterious domain demanded the utmost from our imagination, for all we had seen of the outer worlds were fuzzy photographs from telescopes. We knew our first pathfinding photographic and scientific flights to those far planets would be very special. They would take us to strange and wonderful environments with enormous diversity and an astonishing amount of geological and atmospheric activity. At first we thought that spacecraft would reveal planets that are fairly placid, with moons that had ancient craters like our own Moon and Mercury. We were in for a surprise.

Among the giant planets we found countless wonders and mysteries, and on their moons we found conditions similar to those before – and perhaps during – the emergence of life on Earth. A dramatic series of mind-expanding vistas greeted our first pioneering ventures to those distant worlds.

When the Apollo astronauts stepped from their excursion module to the dusty lunar surface, scientists and engineers were already working on concepts that would result in the relatively modest 1972/73 Pioneer flights to Jupiter and Saturn and the two Voyager spacecraft, which flew past all the outer planets but Pluto. From these projects in turn would come Galileo, a long-lived Jovian orbiter, and Cassini, an equally ambitious odyssey to Saturn.

Even in the midst of the Vietnam War, spaceflight planners were hard at work, notably at JPL, a remarkable institution managed for NASA by the California Institute of Technology. Here in the late 1960s, in an immaculately landscaped collection of laboratories and high-rise buildings in the hills above Pasadena, a crack team of scientists and engineers was charged by NASA with developing a Thermoelectric Outer Planets Spacecraft (TOPS), also known as the 'Grand Tour' spacecraft. While that particular mission never flew because of budget restrictions, the study provided much valuable information for the missions that were to follow. From its ashes grew the Jupiter/Saturn '77 Mission, approved by Congress in 1972 and later renamed Voyager.

Now as then, the birth of a new NASA project is an involved and expensive process. Let's take an example. Europa, one of Jupiter's great Galilean moons, became the focus of considerable interest and excitement following the discovery of a possible water ocean beneath its icy crust. A Europa Orbiter mission was proposed for 2003, and a 'Science Definition Team' formed to work with counterparts at JPL. The team solicited and received reports and recommendations on how to go about the job from universities, government laboratories, and industry; chairman Chris Chyba briefed NASA's Solar System Exploration Subcommittee and the National Academy of Science's Committee on Planetary and Lunar Exploration on their progress and gathered feedback.

In April 1998, the team framed a letter to NASA Headquarters, describing their science objectives, the kind of payload needed to fulfil them, and a plan for engaging the rest of the science community in the project. The science package would have to weigh no more than 5.5 kilos (12 lb), including radiation shielding, and use no more power than a couple of household light bulbs. Yet it would have to determine the presence or absence of a subsurface ocean; map the distribution of any subsurface liquid water and its overlying ice layers; identify candidate sites for future lander missions; characterize the surface composition (especially 'prebiotic' compounds that could be the forerunners or simple life); and describe the moon's radiation environment.

It took a lot of money for the Science Definition Team to do its work, but it would take many millions more to complete the project – money that competitors would be fighting for. Why go to all this trouble and expense? I was struck by the power of two sentences in the team's letter: 'The search for an ocean on Europa is in NASA's greatest traditions of compelling exploration, and presents unprecedented technical challenges. Should the 2003 Orbiter demonstrate the existence of an ocean, Europa will become a primary focus for the 21st century's astrobiological exploration of our Solar System.'

Alas, the project lost its funding in 2002, but backers kept fighting for a 2008 launch. Budget crunches ebb and wane, but spaceflight is here to stay. Twenty eight NASA space science missions should be operating by the end of 2003. As John Naugle, then a deputy NASA administrator, told the U.S. House of Representatives more than two decades ago, there is 'a desire, felt by scientists and nonscientists alike, to discover where we are in the universe, how we got here, and where we are likely to be headed – a quest for a kind of cosmic perspective for mankind.'

Arthur C. Clarke

Humans and Machines,
Time and Space

Creatures of flesh and blood such as ourselves can explore space and win control over infinitesimal fractions of it. But only creatures of metal and plastic can ever really conquer it, as indeed they have already started to do. The tiny brains of our current spacecraft barely hint at the mechanical intelligences that will one day be launched at the stars.

It may well be that only beyond Earth's orbit, confronted with environments fiercer and more complex than any to be found upon this planet, will intelligence be able to reach its fullest stature. Like other qualities, intelligence is developed by struggle and conflict; in the ages to come, the dullards may remain on placid Earth, and real genius will flourish only in Space – the realm of the machine, not of flesh and blood.

But we, and the rest of the solar system, have time to spare. It appears that the past duration of the galaxy is a mere flicker of time, compared to the aeons that may lie ahead. At their present lavish rate of radiation, stars such as the Sun can continue to burn for billions of years; then, after various internal vicissitudes, they settle down to a more modest mode of existence as dwarf stars. The reformed stellar spendthrifts can then shine steadily for periods of time measured not in billions but in trillions of years. The planets of such stars, if at the same distance from their primary as Earth (or even Mercury) would be frozen at temperatures hundreds of degrees below zero. But by the time we are considering, natural or artificial planets could have moved sunward to huddle against the oncoming Ice Age as, long ago, our savage ancestors must have gathered around their fires to protect themselves from the creatures of the night...

By that time, if our species still exists, we may have come to regard super-intelligent machines as our real, albeit non-biological, children. We can conceive of this now, in an abstract form, for machines are undeniably the offspring of human intellect. But as time goes on, the concept of the integrated man-machine family will almost certainly take on the mantle of reality.

The ruin of the universe is so inconceivably far ahead that it can never be any direct concern of our species. Or, perhaps, of any species that now exists, anywhere in the spinning whirlpool of stars we call the Milky Way...

One thing seems certain. Our galaxy is now in the brief springtime of its life – a springtime made glorious by such brilliant blue-white stars as Vega and

Sirius, and, on a more humble scale, our own Sun. Not until all these have flamed through their incandescent youth, in a few fleeting billions of years, will the real history of the universe begin.

It will be a history illuminated only by the reds and infra-reds of dully-glowing stars that would be almost invisible to our eyes; yet the sombre hues of that all-but-eternal universe may be full of colour and beauty to whatever strange beings have adapted to it. They will know that before them lie, not the millions of years in which we measure the eras of geology, nor the billions of years which span the past lives of the stars, but years to be counted literally in trillions.

They will have time enough, in those endless aeons, to attempt all things, and to gather all knowledge. They will not be like gods, because no gods imagined by our minds have ever possessed the powers they will command. But for all that, they may envy us, basking in the bright afterglow of Creation, for we knew the Universe when it was young.

[This essay is extracted in large part from the book *Profiles of the Future* (New York, Holt, Rinehart and Winston, 1984)]

Technology, Dreams, and Little Green Men

'If astronomy teaches anything, it teaches that man is but a detail in the evolution of the universe . . . He learns that though he will probably never find his double anywhere, he is destined to discover any number of cousins scattered through space.'

Percival Lowell

Most of our knowledge of this marvel-filled universe has come from astronomy, telescopes, and robotic spaceflight. It is impossible to think of anything that more exquisitely embodies the technical genius of humankind, in so small a package, as the interplanetary spacecraft. Even Voyager, using relatively old technology in the 1970s, sent enough information back to Earth to encode six thousand complete sets of the *Encyclopædia Britannica*. It sent us marvellous, unparalleled views of Jupiter and its moons – but comparing these images with those that came later from the Galileo spacecraft was like comparing one's view of a book from the top of the Empire State Building with the view of that same book held in one's own hands. 'God clones Himself in Man,' Ray Bradbury wrote in 1980. 'Man clones himself in machines. Machines, if properly built, can carry our most fragile dreams through a million light years of travel without breakage.'

People have always been fascinated by the prospect of visiting other worlds, and by the prospect of finding life on those worlds. Our unstoppable curiosity, the irresistible quest to explore and discover, have fed human imagination throughout history. For some, spaceflight offers the ultimate outlet for these basic human urges. For all of us, it serves as a driver for innovative science and engineering that in turn stimulates economic growth – in consumer products, international economic competitiveness, jobs, and the rest.

It will be useful to have human hands and eyes in space, especially for tasks that require unusual powers of observation and judgement, such as searching for signs of past or present life. The fact that our human brothers and sisters are simply 'out there' also may expand our species' psychological horizons, bringing people closer, with a greater sense of community, than when they are preoccupied with the social and political issues of our small planet.

Still, it can hardly be said that people have a limitless love affair with human spaceflight. When construction of the International Space Station (ISS) began in 1998, it was attended by heated debate over its multi-billion dollar budget. Remember the continuing hubbub over the utility of the Space Shuttle, and even the much-vaunted Project Apollo, the cost of which the eminent physicist Freeman Dyson said 'set back the development of space science by twenty years.' In simple terms, the project was financially unsustainable: even the mighty United States could ill afford to keep it going.

NASA's explanation that the gravity-free ISS would provide an excellent environment for producing more perfect, homogeneous materials – foams, alloys, and crystals – didn't carry much weight with the ordinary citizen. The story made great television, however, and viewers watched avidly as the first pieces of this 460-tonne engineering marvel were manoeuvred into place.

Not much was made of the project's likely contribution to human spaceflight. Yet our first mission to another planet will almost certainly be assembled in Earth orbit. Mistakes and other lessons learned from the space station project will contribute to its success, just as the errors of early robotic missions will help later flight architects assure the success of crewed vehicles when they reach their targets.

Sometime in the future, we can foresee that we will be assembling a great space cruiser in orbit. Imagine the largest airliner fuselage you can, and stretch it into a circle as far across as a football field. Armour it with lightweight, radiation-resistant materials. Inside the hollow wheel, install control decks, laboratories, crew quarters, and salons for exercise, recreation, and dining. Design a propulsion and attitude control system and put it in the centre of the circle. Connect this hub to the rim with four strong spokes that hold the whole affair together and provide access corridors. Rotate the craft, simulating gravity so that the physiological effects of weightlessness will be reduced or even eradicated. Now fill the reservoirs between the spokes with hydrogen propellant. Start the reactor, load the crew, fire the rockets to break out of Earth orbit, and we're off to the outer planets!

The dream will come true, someday, but it will take many decades of development. First we must undertake human exploration of the Earth-Mars environment. We have three moons to explore out there, and the Red Planet itself. For its part, the American Institute of Aeronautics and Astronautics (AIAA) is recommending that the U.S. 'lead a co-operative international effort

The International Space Station captured by the Space Shuttle in December 2000. Information on the effects of zero-gravity conditions aboard the ISS, and other astronautical programmes, supports planning for future human exploration of the planets.

to conduct further human exploration missions to the Moon and subsequently to Mars in the early- to mid-twenty-first century.' Japan's National Space Development Agency is planning Earth orbiters and eventually an expedition to build an astronomical observatory on the Moon. China, which orbited a capsule capable of carrying astronauts in late 1999, is interested in landing its own astronauts on the Moon, and the European Space Agency would like to establish a permanent Moon base, with a human landing in about 2030.

The success of human lunar exploration depends largely on how clever we are in processing lunar soil, known as regolith. This is because as much as 45 per cent of regolith is composed of that vital element, oxygen. 'Lunox' could be used for generating air and water, as well as powering liquid-oxygen/liquid hydrogen rockets, since this fuel is 85 per cent oxygen. Or it could be used to supercharge the hydrogen propellant used in advanced nuclear thermal rocket engines (see pp.108–110).

After we have successful bases on the Moon and Mars, it will be time to start designing our human-crewed flight to Jupiter and its moons. These are among the distant goals of NASA's Human Exploration and Development of Space enterprise, whose mission is 'to open the space frontier by exploring, using, and enabling the development of space and to expand the human experience into the far reaches of space.'

The least-known factors in human interplanetary spaceflight concern the astronauts themselves. It's possible that space cruisers capable of taking astronauts to Mars could have been ready to fly in the 1970s, had the programme been given the necessary priority. But a flight to Mars and back, or an extended stay on the Moon or Mars, would impose daunting physiological and psychological burdens on the crew. The problems of weightlessness and confinement have not yet been defined, but we know that they are considerable, and that solving them will make the enterprise extremely expensive – or extremely risky (see pp.168–171).

In interplanetary terms, Mars would represent a very short trip indeed, though a round-trip visit will require a life-support system that operates for 20 to 30 months without failure. On the other hand, human excursions to the outer planets using current technology would take up most of an astronaut's career. Even if we can design workable life-support 'closed systems' – in which food and oxygen would perhaps be provided by on-board mini-farms – massive human problems would remain.

'In my heart I know we will not fully experience Mars and get all the science we can until a human can get there and lift a rock or drill down,' NASA

Stephen Braham Simon Fraser University
Pascal Lee SETI Institute and NASA Ames Research Center

Bases on the Moon, Mars and Beyond

In-depth exploration of planetary bodies such as the Moon and Mars will require more than robotic missions – or even human visits if they are of short duration. It will require the establishment of actual bases where personnel and resources can provide sufficient logistical support and on-site scientific research facilities to enable a safe and effective long-term exploration capability.

The exploration of our own planet over past centuries has seen the use of many forms of such bases. For example, the exploration of Antarctica developed as we moved from initial reconnaissance by ships ('portable bases') to the building of semi-permanent bases, to the establishment of the fixed scientific research stations that have made modern Antarctic research so effective.

In the Canadian Arctic we ourselves experienced the value of establishing even a semi-permanent base to adequately support field exploration activities under difficult and remote circumstances. Similarly, bases on the Moon and Mars are expected to serve as outposts from which in-depth exploration can be undertaken effectively.

A common driver in the establishment of a base or remote outpost, be it in Antarctica or on a planetary body, is the need to put together a safe haven and adequate logistical infrastructure in a distant, hard-to-access place where there was initially none. Though establishing a base is inherently a costly enterprise, it should be viewed as an essential investment – indeed, the first infrastructural objective – if substantial exploration activities are anticipated. Scenarios proposing that each new crew arriving on a planet set up camp some distance from previous camps make little sense if long-term, in-depth exploration is contemplated.

While Arctic and Antarctic environments are lethal if adequate equipment and gear are not available, the environments on the Moon and on Mars are immensely more challenging and will be lethal to unprotected humans virtually instantaneously. Nevertheless, many valuable lessons about living in terrestrial polar stations apply directly to living on another planet. The psychological factors associated with isolation and distance from home come to mind immediately, but other aspects such as safety procedures for the conduct of outdoor exploration activities (buddy systems, radio calls, establishing caches, etc.) provide many opportunities for relevant studies.

Thus, an important step in learning how to live and work in future habitats and bases on the Moon, Mars and beyond is to experiment with the use of similar facilities – a field base camp or a simulated habitat – on our own planet. On Devon Island, in the Canadian Arctic, we have been investigating operational issues that will help optimize the conduct of field science and exploration activities on Mars. One critical requirement, beyond conducting safe and productive exploration activities, is to convey adequate awareness of the situation and all the finer details arising out of daily exploration to others 'back on Earth.' While they need a measure of autonomy because of the large distance separating them from immediate help from Earth, bases will need to remain in close contact in order to optimize safety, scientific yield, and the sharing of the excitement of exploration with all humans.

Being closely networked with Earth has its own special problems. Because of the great distance involved, communicating with Mars will be more difficult than communicating with the Moon. Not only does it require more powerful signals, but the distance between the two planets imposes a time barrier ranging from 4 to 20 minutes, the one-way transit time for radio signals travelling at the speed of light. Thus, a base on Mars will be inherently more isolated from the Earth than a base on the Moon, and therefore will need to be more autonomous.

The time delay will also affect daily exploration of the planet. On the Moon, during moonwalks, it was possible to send live video and audio to Earth and, within seconds, have mission controllers and scientists radio commands and feedback to the crew. On Mars, we can still send live video and audio to Earth, but Marswalking astronauts may have to move on to their next destinations before receiving comments. They will have to respond by returning later to important locations noticed by Earth-based researchers, either on their way back to their Mars Habitat, or, in the worse case, on a separate Marswalk (or Marsdrive). Part of developing analogue Mars exploration on Earth involves understanding how we can structure these exploration outings to get the most out of the proverbial planetary exploration buck.

In addition to the substance of daily exploration, field scientists exploring the solar system will need to concentrate on the bread and butter needs of surviving far from home. On Mars, if something goes wrong, astronauts will often have to react before Earth even receives a trouble signal. Indeed, crews on the Russian space station Mir, and the early stages of the International Space

Station Alpha, have already learned to do this because early communication systems to these stations allowed only sporadic communication, often with considerably more delay in response than the systems we can expect on a Mars or Lunar mission! Thus these crews have made, and, with Alpha, continue to make, major strides in solar system exploration, while other scientists are also helping design future planetary bases in Earthbound simulated facilities such as the one on Devon Island.

An artist's impression of a manned base on Mars. This base has been modelled on similar bases in the Antarctic.

administrator Ed Weiler, a space robotics advocate, told science writer Andrew Lawler. 'Even more important to the human soul is seeing Mars with human eyes. That is worth something that is not part of being a scientist, but part of being human.'

Planets, Asteroids, Comets, Mystery

Wherever there is a mystery, people have always found challenge, even if the solutions to those mysteries do little to change the material quality of our lives. Scientifically, we humans want to learn more about the form and origin of the solar system; to gather information that will help us unify the laws of physics; to gain new insight into the origins of life. Philosophically, we want to use these findings to learn more about ourselves and the meaning of our existence. Economically and politically, we look for clues to the future of our own planet, and to the possibility that the riches of other planets may be used to supplant the gradually dwindling natural resources of Earth.

We are gradually accumulating a store of exciting new information as we continue to make direct measurements of surface conditions, as we monitor the planets' internal and atmospheric makeup, and as we learn more about their magnetic fields and those of interplanetary and interstellar space. As a corollary to learning more about our planetary neighbours, we are gaining insight into the nature of planetary systems that exist around stars other than our Sun, and making better-educated guesses as to whether Earth-like (or at least habitable) planets exist in other parts of the Universe.

There are several kinds of planets in the astronomer's vocabulary. First are the nine major planets, most with moons and/or rings. Additionally there are countless minor planets – more commonly called asteroids – ranging from around 930 kilometres (580 miles) in diameter down to a kilometre or less. More than 10,000 of these have been identified, mostly in the asteroid belt between Mars and Jupiter. Finally there are the comets, dark chunks of ice and dust that circle the Sun, often in highly elliptical orbits.

The major planets are categorized in several groups. The terrestrial (or Earth-like) planets include Mercury, Venus, Earth, and Mars; the giant (or Jovian) planets are Jupiter, Saturn, Uranus, and Neptune. Pluto is called either terrestrial for its size, or Jovian for its position: but in either case it is one of the outer planets along with the giants. Additionally, Venus and Mercury are sometimes called inferior planets because their orbits lie within that of Earth.

The Jovian planets are also called 'gas giants' because they are composed mostly of hydrogen and helium.

The terrestrial planets are small, ranging from 4,880 to 12,750 kilometres (3,030 to 7,920 miles) in diameter; they rotate in 24 hours or more; and they have relatively thin atmospheres. The giant planets, by contrast, range from 49,500 to 143,000 kilometres (30,700 to 88,800 miles) in diameter; rotate in between 9 to 17 hours and have very thick atmospheres. The terrestrial planets are much more dense and rotate more slowly than the giant planets, and, for these reasons, they do not bulge so much at their equators as they spin.

The distance of each planet from the Sun, and its movement around its elliptical orbit, is explained by the laws of planetary motion first discovered by Johannes Kepler. These show that the planets move in ellipses rather than perfect circles, and that they move faster in their orbits when they are closer to the Sun. To remember just where the orbit of each planet lies, it's helpful to remember the mnemonic, 'Martha Visits Every Monday, Just Stays Until Noon, Period.' Not only are the first letters of these words the same as those of the planets, in sequence outward from the Sun – but if you remember the punctuation, the commas separate them into the terrestrial planets, the giant planets, and that oddball of them all, Pluto.

Countless millions of asteroids are concentrated in the asteroid belt, a 300-million-kilometre (186-million mile) wide ring between Mars and Jupiter. The material in the belts travels at about 20 kilometres per second (12.5 miles per second) and ranges in size from dust particles to Alaska-sized behemoths.

When asteroids break up, as a result of occasional collisions, the pieces may become smaller asteroids or meteoroids, often falling into more elliptical orbits that cross the paths of the inner planets. If the meteoroids enter Earth's atmosphere, they are visible as meteors (commonly called shooting stars). When the fragments survive their trip through Earth's atmosphere and land on the ground, they are known as meteorites. It has been estimated that a billion tonnes of rocks have also been blasted free of Mars' weak gravity by meteorite impacts, and made their way to Earth since the two planets formed. One of the most amazing meteorites ever, a 250-tonne monster that slammed into frozen Tagish Lake in northern Canada in January 2000, is believed to contain dust grains from the supernova that seeded the infant solar system.

When dust grains are eroded away from asteroids and go into their own orbits around the Sun, they become part of the zodiacal cloud. As this faint sheet of dust catches the sunlight, it appears as a hazy glow around the ecliptic (the plane of the solar system) after sunset or before sunrise.

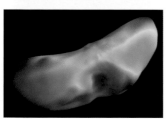

Images of the asteroid Eros from NASA's NEAR Mission. The spectacular view of the north polar region (top) was constructed from six images taken from an orbital altitude of about 200 kilometres (124 miles). This vantage point highlights the major features of the northern hemisphere: the saddle seen at the bottom; the 5.3-kilometre (3.3-mile) diameter crater at the top; and a major ridge system running between the two features that spans at least one-third of the asteroid's circumference.

Eros' density and spin combine to create a bizarre pattern of what is 'uphill' and 'downhill'. In this view (bottom), a map of 'gravitational topography' has been painted onto a shape model. Red areas are 'uphill' and blue areas are 'downhill'. A ball dropped onto one of the red spots would try to roll across the nearest green area to the nearest blue area.

Even asteroids come in categories. In and around Jupiter's orbit there are several hundred asteroids known as Trojans; Near Earth Asteroids are called Amors, Atens, or Apollos depending on their orbits.

Some asteroids appear to be little more than pockmarked irregular chunks of rock. But when examined in 1995 the Arizona-sized, football-shaped asteroid Vesta turned out to be so unusual and varied in composition that Georgia Southern University researchers wanted to record it as a new terrestrial planet. And for good reason – since Vesta has an exposed mantle and ancient lava flows, and was roughed up by collisions that likely sent bits and pieces hurtling to Earth as meteorites. The following year, Hubble's Wiffpick confirmed that there is also a giant impact crater on Vesta, representing two million cubic kilometres (half a million cubic miles) of rock, blasted away by an ancient collision with another asteroid.

NASA's NEAR (Near Earth Asteroid Rendezvous) Mission swung by asteroid Mathilde in the late 1990s, then headed for Eros, an asteroid that measures about 15 x 15 x 40 kilometres (9 x 9 x 25 miles). This was more than a rendezvous – NEAR didn't stop at producing ultra-close-up images. In early 2001, it performed the spectacular feat of actually landing on the potato-shaped object, almost exactly 200 years after Giuseppe Piazzi had discovered the first asteroid, Ceres. (The largest body in the asteroid belt, Ceres accounts for more than one-third of the estimated total mass of all the asteroids.)

And then there's the question of when another big object will strike Earth. Ninety per cent of such impacts are thought to come from asteroids, while the rest are from comets. A number of organizations are attempting to locate and track Near Earth Objects (NEOs) which might represent a threat to Earth. These include the Safeguard Survey run by NASA and the U.S. Air Force, and the internationally funded Spaceguard Foundation. (The name 'Spaceguard' comes from Arthur C. Clarke's book *Rendezvous with Rama*, and is often applied to the entire search for NEOs). It is estimated that there's a one in 20,000 chance of you or me being killed as the result of a NEO impact – a small risk, but enough to merit keeping a lookout, since although rare, such impacts are inevitable in the long term and their effects are devastating. Its one reason of many to support the idea of establishing human colonies elsewhere in the solar system.

Are asteroids simply good-for-nothing bits of debris, hovering threats to Earth's well-being? Not entirely. The idea that someday we will mine asteroids for ice and precious minerals is a subject of continuing discussion. The value of the metals in one sample asteroid, a 2-kilometre (1.4-mile) rock called Amun, is valued at a mind-boggling $20 trillion!

David Morrison
Senior Scientist, NASA Astrobiology Institute, Ames Research Center

Killer Asteroids

The most severe known natural hazard is impact from a comet or asteroid. Unlike most hazards, such as earthquakes, storms, or volcanic eruptions, there is no natural upper limit to the size and energy of an impact. A large impact could inflict great damage upon global civilization or even perhaps on the human species. Of course such impacts are exceedingly rare. We do not expect them to happen within our lifetimes or even the next several centuries, but we cannot exclude the possibility.

Most scientists first became aware of the continuing impact hazard in the 1980s, after the extinction of the dinosaurs 65 million years ago was linked to the Chicxulub impact. The cause of the extinction was first revealed by the chemical signature (from presence of iridium and other rare metals) of extraterrestrial material at the boundary between the Cretaceous and Tertiary periods of history. Later this material was identified with a 15-kilometre (9-mile) asteroid (or perhaps a comet) striking in Mexico's Yucatán peninsula and creating a crater 200 kilometres (125 miles) across. Examples of smaller but more recent impacts include the iron asteroid that formed Meteor Crater in Arizona and the rocky asteroid that exploded over the Tunguska River in Siberia in 1908. Both of these projectiles were about the size of a 12-storey building and had explosive power equivalent to the largest nuclear weapons.

In 1991 the U.S. Congress asked NASA to evaluate the current impact hazard and suggest ways to deal with the problem. The U.S. House of Representatives Science Committee wrote that 'the detection rate must be increased substantially, and the means to destroy or alter the orbits of asteroids when they do threaten collisions should be defined and agreed upon internationally. The chances of the Earth being struck by a large asteroid are extremely small, but because the consequences of such a collision are extremely large, the Committee believe it is only prudent to assess the nature of the threat and prepare to deal with it.' This remains an excellent summary of the issue.

Scientists have concluded that the greatest hazard is from Near Earth Asteroids (NEAs) larger than one kilometre (0.6 miles). The biggest damage comes not from the blast itself, but from the global pall of stratospheric dust that would follow. This dust would darken the sky, causing temperatures to drop and killing most crops. It also seems clear that if such a collision were predicted far enough in advance, we could develop the technology to deflect or destroy the threatening asteroid. Impacts represent not only the worst imaginable natural catastrophe, but also the only one against which we might be able to effectively defend ourselves. This consideration led to the recommendation that we begin a concerted effort to find all of the NEAs larger than 1 kilometre in diameter (there are estimated to be about a thousand) and determine their orbits in order to identify any that threaten collision with the Earth. This search for NEAs is called the Spaceguard Survey.

When the Spaceguard Survey was first proposed, astronomers had discovered fewer than 100 of the estimated 1000 NEAs larger than 1 kilometre. Most of the discoveries came from photographic patrols carried out with a handful of small telescopes. The photographic images were scanned by hand to find the rare image of an asteroid moving against the background of millions of star images. The result was approximately one new 1-kilometre NEA discovered per month. The total effort of the citizens of our planet directed toward protecting us from a catastrophe that could kill more than a billion people was less than the staff of an average McDonald's restaurant.

To accelerate the search, NASA funded astronomers to develop automated telescopes equipped with many CCD detectors to image the sky. Computers compared the images to identify the moving asteroids. New software also made it possible to project the orbits of NEAs forward in time with high precision. Dedicated telescopes began work at Kitt Peak in Arizona and on the summit of Haleakala in Hawaii. By the late 1990s, the Spaceguard Survey was dominated by two small one-metre (40-in)) U.S. Air Force telescopes (called LINEAR) in New Mexico, operated with NASA funds. The discovery rate rose from one a month to more than 10 a month, pointing toward cataloguing 90 per cent of the NEAs larger than one kilometre by 2008. By mid-2002 we had bagged almost 600.

So far no NEA has been found on a collision course with our planet. But we need to think about what we would do if an impact were predicted. Given decades of warning from the Spaceguard Survey, we would probably send a series of spacecraft to the asteroid, first to study it scientifically ('know your enemy'), then to change its orbit (probably with nuclear explosives). The NASA NEAR-Shoemaker mission demonstrated the technology to fly to an asteroid, orbit it, and even land on the surface. What we have not yet done is undertake any serious studies of deflection.

Carl Sagan once pointed out that from a long-term perspective, one of the requirements for survival of a species was to learn to control our impact environment, and hence to become a spacefaring civilization. Species that do not find ways to protect themselves from large impacts will die – sooner or later, depending on luck.

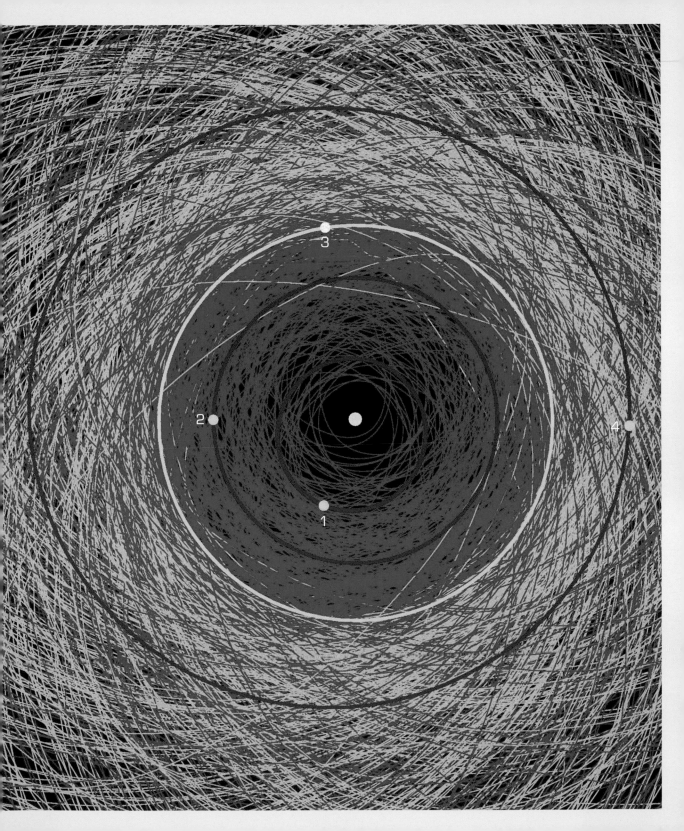

The nucleus of Halley's Comet as seen by the ESA's Giotto spacecraft in 1986. The dark nucleus is clearly visible against the bright coma. Several surface features are visible, including a 'crater' between the two bright spots on the left edge and a sunlit 'mountain' on the right side. The light 'jets' are material sublimated from the surface of the comet in the direction of the Sun. At its closest encounter Giotto passed within 600 kilometres (400 miles) of the nucleus.

Comets are another phenomenon of the solar system – known for their spectacular luminous tails, extending up to 150,000,000 miles across space. Unlike the planets, comets usually trace a highly elliptical, elongated, orbit around the Sun. Halley's Comet, for example, follows a path that extends beyond the orbit of Neptune at its farthest distance from the Sun, but comes within the orbit of Venus at its closest approach, once every 76 years.

In 1986, ESA's Giotto probe sidled up beside Halley's comet and returned amazing images of a dark, potato-shaped object, with crater-like hollows and mountains, measuring only 15 kilometres (9 miles) long and half as wide. This was Halley's nucleus, the source of all its activity. Its very low density suggested a porous texture, and it appeared to be coated with a thick layer of dust.

Comets are often nicknamed 'dirty snowballs' because their nuclei consist of trapped dust and ice. When they get near the Sun, the ice evaporates and makes a bright 'coma' that trails back into the familiar comet's tail as it is swept by the solar wind. While the comet's nucleus may be only a few kilometres across, its coma may be many times the diameter of Earth. The largest comas may reach the size of the Sun!

Once seen as objects of fear and ill omens, comets today are still admired for their spectacle. But they also attract great scientific interest because they are pristine remnants of the early solar nebula, and are thought to have brought water to the inner planets bearers in the early days of the solar system.

Close relatives to the comets are the mysterious, icy objects called Centaurs, suspected refugees from the Kuiper Belt that are found in the realm of the giant planets. These include the cratered Asbolus, 80 kilometres (48 miles) wide, orbiting between the paths of Saturn and Uranus. (See pp.250–252.)

Celestial Timetables: The Ephemerides

Planning for every interplanetary space flight begins with a study of three orbits around the Sun – those of Earth, the target body (planet, asteroid, etc.), and the spacecraft. The velocity and direction of these bodies must be known with great precision, for ballistic flight paths may be altered only slightly after the initial insertion into orbit.

In a sense, the space flight begins before lift-off, for the spacecraft is already orbiting the Sun as part of Earth, and even the location of launch sites is designed to take advantage of the speed boost provided by Earth's rotation. At this point, the important task is to select a suitable launch date – say one that will start the spacecraft on the fastest-possible flight to a target planet. To do this, planners must predict the future positions of Earth and the target planet so that they know when the two will approach closest to each other. Then, working back from this date, they select the best launch opportunity for a spacecraft travelling a given flight path at a given velocity. Within this time frame there are daily launch windows during which the spacecraft is ideally 'aimed' on its trajectory by the natural rotation of the Earth.

Once the spacecraft has separated from its launch system, precise targeting becomes increasingly more critical. Its trajectory must intersect that of the planet at exactly the right time and place, with minimum use of the limited course-changing power provided by its thrusters. Otherwise the mission will fail or, at best, be unable to properly fulfil its programmed tasks.

'It's roughly like trying to shoot a duck with a rifle,' explained one NASA mission engineer. 'You can aim your gun wherever you want – but that doesn't do you much good unless you can figure where the duck will be when the bullet gets there.'

'And,' added Charles Kohlhase, mission analysis and engineering manager for the Voyager project, 'if the route is using gravity assist by other bodies first, then it's like aiming your rifle to ricochet off several stones first, before hitting the duck!'

'I used to worry a lot about the propellant situation,' Kohlhase said. 'Would there be enough? I frequently dreamed about this, often awaking before dawn to reconsider the equations and assumptions we used to estimate the delta-vee (overall need for changes in velocity). Jupiter was so massive and had such enormous potential to deflect the trajectory, yet the computed delta-vee requirements kept turning out to be reasonably small. Was there an error lurking in there somewhere, or were the navigation dispersions such a tiny fraction of the swingby corridor width and distance from Jupiter that it simply was not a concern?'

We can accurately plot the future path of the Earth around the Sun, and we can control the trajectory of interplanetary spacecraft with a certain predictable precision. Forecasting the orbits of other planets, however, requires such special expertise that it has become a specialized pursuit within the field of astrophysics.

These forecasts are put together in the form of ephemerides – astronomical timetables that originally were computed with the help of telescopic observations of the sky.

Ephemerides are traditionally used by navigators to chart positions at sea, and by astronomers to locate the 'targets' for their telescopes. Information is given as it relates to the celestial sphere (described below), using a system which roughly parallels geographical reference points on Earth, where we use longitude – the pole-to-pole delineators measured from the zero point or 'prime meridian' running through at the Royal Observatory in Greenwich, England – and latitude, the horizontal lines that reach around the Earth parallel to the equator. The Martian prime meridian was defined by the German astronomers Wilhelm Beer (1797–1850) and Johann H. von Mädler (1794–1874) around 1830, and used by Giovanni Schiaparelli (1835–1910) in his 1877 map of Mars – some time before the 1884 conference which finally fixed our own planet's meridian.

To understand the celestial sphere, consider the stars as projected on the inner surface of a hollow sphere. The 'North Star' Polaris appears approximately at the North Pole of this sphere; the faint star sigma Octantis near the South Pole. The celestial centre is the Earth's centre; its equator is parallel with the Earth's. The Sun circles the sphere in a path – the ecliptic – which passes north and south of the equator at the vernal and autumnal equinoxes, respectively.

Celestial longitude – the lines from pole to pole – is known as right ascension and is measured in hours, minutes, and seconds eastward from the vernal equinox point. Twenty-four hours represents the entire circumference, as it would in the case of terrestrial time zones.

Celestial latitude, on the other hand, is termed declination. Its lines of reference are measured in degrees, from zero at the celestial equator to 90 or -90 at the poles.

As an example of how these coordinates are used, let's take the position of Sirius, an important star for navigation since it is so very bright and easy to detect. In the shorthand of the ephemerides, Sirius' position would be described as R.A. 6h 43m, Dec. -16° 39'. In other words, it is 6 hours and 43 minutes east of the vernal equinox and 16 degrees 39 minutes south of the celestial equator.

Bode's Law

Actual distance (A.U.)

0.39   0.72   1   1.52   2.77   5.2   9.54   19.19   30.1   39.5

asteroid belt

0.4   0.7   1   1.6   2.8   5.2   10   19.6   —   38.8

Bode's Law prediction (A.U.)
(1 A.U. = 93,000,000 miles)

Astronomers started looking for the minor planets in earnest after Johann Elert Bode (1747–1826) popularized a strange mathematical formula devised in 1776 by a German compatriot, Johann Daniel Titius (1729–1796). Generally known simply as Bode's law, this peculiar calculation gives surprisingly accurate clues to where one might find planets.

Titius knew the distances from the Sun of Mercury, Venus, Earth, Mars, Jupiter, and Saturn, and went to work on a theory that might be some mathematical relationship between them. He assigned zero (0) to the Sun, 3 to Mercury, and doubled the number for each planet thereafter. Then he added 4 to each number and divided the sum by 10. And presto! – there were the distances, neatly given in astronomical units.

The orbit between Mars and Jupiter (2.8 A.U.) carried no obvious planet, so the astronomers looked closer, and there was the asteroid belt, which could be described as broken planet fragments, or simply as a 'failed planet.'

Only Neptune failed to comply with Bode's Law. And nobody, so far, knows why Bode's Law works so well. The odds are that it is an accidental artefact of early scientific enthusiasm.

The 'orrery' solar system model was a popular instrument of astronomical instruction in the 18th and 19th centuries. Here, only the six planets closest to the sun are represented. Uranus, Neptune and Pluto were discovered in 1781, 1846 and 1930 respectively. The frame around this device represents the 'celestial sphere'.

These may be regarded as unchanging coordinates, since we have agreed that the stars are virtually motionless. Because they move against the backdrop of stars, however, planets and spacecraft coordinates are constantly changing. Thus we must add the element of time to pin down when the body of interest will be in the position we describe. Twenty-four-hour Universal Time, which is the same as the solar time at the zero (Greenwich) meridian, is used in the ephemerides as well as in other scientific measurements. It is the same all over the world, regardless of the location of the observer.

The first modern ephemerides were published in 1679 by the French cleric Jean Picard (1620–1682) under the title *La Connaissance des Temps* (*The Knowledge of Times*) They were designed at a time when measuring longitude on Earth was a major problem. The best way to do this would be to know the accurate time as measured at your point of departure, and compare it with the local time measured from the Sun. Since accurate seagoing clocks were nonexistent, it seemed logical to use the one clock everyone had access to – the solar system. If the positions of planets or even the moons of Jupiter could be measured and compared to a table of their predicted positions, the 'Universal Time' according to the heavens could be found and compared to local time. However, Picard's ephemerides were not quite up to the job, and it was Britain's Astronomer Royal, Sir Nevil Maskelyne (1732–1811) who later produced the first truly practical ephemerides with the aid of data compiled by Johann Tobias Mayer (1723–1762) of Göttingen, Germany.

It was not long before Sir Charles Boyle, 4th Earl of Orrery (1676–1731), had put a young technician to work on a machine that was to be the ancestor of the modern planetarium. It used a system of brass balls and clockwork to show the motion of bodies in the solar system, and was named the 'orrery' in honour of its patron. We think of orreries today as quaint antique conversation pieces, but they live on in computerized form. In the 1980s, Massachusetts Institute of Technology developed a computer-driven Digital Orrery that could analyse the motion of ten gravitationally interacting bodies (such as the Sun, planets and moons). This remarkable machine was able to predict the orbits of the outer planets 9,845 years into the future.

During the 1800s, national ephemerides were published by many nations. The complexity of compiling them in those pre-computer days inevitably led to a programme of international cooperation. Collaboration between the directors of various national ephemerides began with a conference in Paris in 1896; today it is somewhat formalized, with several nations specializing in the ephemerides of certain groups of celestial bodies.

These tables are quite adequate for Earth-based astronomy and for navigation at sea and in the air. But problems arise when scientists attempt to relate them to space navigation needs.

The classical ephemerides, as published in the *American Ephemeris* and the *British Nautical Almanac*, have an accuracy of between 0.5 and 2 arc seconds along the direction of the planet's motion (an arc second is 1/3600th of a degree. Whereas an arc second is a relatively short distance at the Moon, the angle naturally covers a larger region of space at greater distances, until it represents, say, around 1000 km (620 miles) at Mars and 8,000 km (5,000 miles) at Saturn.

Beginning in 1966, JPL scientists launched an all-out effort to improve upon the classical ephemerides. Their job: to combine reliable optically recorded information with data from spacecraft tracking and measurements made by bouncing radar beams from the planets.

JPL's computer experts were given an incredibly large collection of astronomical information to translate into data NASA could use for space navigation. It involved the combining and processing of two general types of information – on the one side some 90 different factors affecting the physical characteristics of the Sun, solar system, planets, and the spacecraft themselves; on the other, reams of observational data gathered by the U.S. Naval Observatory (since 1910), the Mariner spacecraft, and a number of radar installations. After this complicated process was completed, our space scientists' knowledge of the planets' positions was more precise than ever before. But the ephemerides for the outer planets were still less accurate than required. And since the effectiveness of radar observations decreases rapidly with distance, further improvement would have to rely on data from probes that were actually sent to the outer planets. Even without these, however, flight plans were almost unbelievably accurate – Voyager's arrival at Neptune, after a journey of 7.1 billion kilometres (4.4 billion miles), was within 100 kilometres (62 miles) of the spot targeted 12 years before!

A planet's gravity can be used to speed up or change the direction of a spacecraft. What the spacecraft gains in velocity, the planet loses. For example, two separate slingshot orbits of Earth increased Galileo's speed by 32,000 kph (19,900 mph) and slowed Earth's orbit by 5 billionths of an inch per year.

Billiards in the Sky

Prior to the early 1960s, a space flight from Earth to Uranus would have involved a tedious, sixteen-year journey. Almost overnight, flight times dropped to a fraction of their former values – to a minimum of six years in the case of Uranus – and outer-planets space flight at last became part of the headlines. One particularly intriguing part of the news was that these missions would be cut short without the need for expensive new propulsion systems. They would be accomplished with relatively simple ballistic trajectories, thrown out into space with a brief initial burst of rocket power.

The breakthrough came with the introduction of 'gravity assist' for interplanetary space flight. The theory was that planetary missions could be made vastly more efficient by deflecting spacecraft with planetary gravitational fields, thus changing the paths they would normally follow around the Sun.

Because it is so massive, Jupiter has an intense gravitational field and thus can be extremely useful in changing the direction of spacecraft even when they are travelling very fast. Jovian gravity makes it possible to cut down both the launch energy required and the time needed to fly to the outer planets – provided, of course, that the planets are 'lined up' so the spacecraft's flight path requires only small corrections to carry it to the proper place at the proper times.

Mariner 10 used gravity assist to skim around Venus for its first interception of Mercury in 1974. Additional gravity assists permitted the craft to have two additional flybys of Mercury. Then in 1979, Pioneer 11 used Jupiter's gravitational field in order to become the first spacecraft to intercept Saturn.

The laws of Kepler and Newton, interlaced with the findings of modern space physicists, provide the tools we need to help plot our spacecraft's journey among the outer planets. But for a simple introduction to the ballistic trajectories that spacecraft follow, let's turn to the so-called gravity pit analogy.

Imagine a thin, pliable membrane stretched tightly a foot or so above the floor. Then attach a thread to the underside of the membrane at the centre and pull it down so that a deep, wide conical pit is formed. This is the Sun in our analogy of the solar system.

Next, attach more threads to the underside of the conical membrane at various points around the centre and pull them down to form a number of smaller pits. These represent the planets.

When the model is completed, the rims of the smaller pits should curve outward at an ever-decreasing angle, until there is a relatively level area between them. The pits – representing gravitational fields – thus are limited and between them are large areas through which an object can move of its own momentum, virtually free of any external force but the Sun's. Take careful aim and shoot a marble across the upper part of the roughly cone-shaped pit and it will roll down into the cavity and out again. It won't pick up any speed, but it will change direction.

We can't take this analogy very far, one reason being because celestial mechanics requires three dimensions plus time. However, it gives a two-dimensional hint of how gravitational fields will affect approaching bodies.

Let's move closer to real life, to something that we can't begin to model with our plastic membrane. As we've seen, the gravitational fields of planets can be used to change the direction of a body that dips into them and out again. The other great advantage of gravitational assist comes from the energy locked up in a planet's movement around the Sun. Inevitably, as a spacecraft encounters a planet's gravitational field, its own movement will be changed relative to the Sun. It will gain or lose energy, depending on the direction of its approach into the planet's gravitational field; in other words, it will appear to speed up or slow down as it returns to its solar orbit.

Imagine a moving walkway that has a speed of 3 kph, and crossing onto that walkway while you're walking at 5 kph, and that you then continue to walk without slowing. By the time you stumble off the moving walkway, you're travelling at 8 kph. (That's 8 kph relative to the floor, but still 5 kph relative to the moving walkway.)

This illustration roughly resembles gravity assist, where spacecraft get a 'free ride' by passing close to another planet or moon. In this way they undergo a predictable change in velocity at the same time that they change direction.

As armchair astronauts dig deeper into how this all works, the confusion factor looms tall. One reason is that we are so accustomed to using the word 'speed' as synonym for 'velocity.' Actually, speed is a 'scalar' quantity, measuring the distance covered in a given time, regardless of direction. Velocity, on the other hand, has both magnitude and direction, and so we say it is a vector quantity.

Sometimes velocity is figured for the spacecraft as it relates to the target planet. Again this involves relative velocities and directions of two bodies. Suppose you see a friend 100 metres away and begin to run directly towards him at 1 metre per second. If he were standing still, you would reach him in

100 seconds. But if he were moving, your relative velocity would depend on his direction and speed of motion as well as your own. Essentially the same effect occurs as a spacecraft approaches its target planet.

So we see that a given spacecraft will have different velocities with respect to the Earth, each of the other planets, and the Sun.

The Grand Tours

From their studies of planetary positions and velocities, JPL scientists were able to determine that spacecraft could have visited all the outer planets before the year 1990. In fact, by mid-1969, they had drawn up initial plans for two Grand Tours to the outer planets. Thanks to the development of radioisotope thermoelectric generators (RTGs), there would be power to send back information about these mysterious planets, from regions so far from the Sun that other energy sources would be utterly useless. But the billion-dollar price tag proved too expensive, and NASA told the planners to think again.

Clearly, there had to be a mission to Jupiter and Saturn in 1976, 1977, or 1978. In each of these years there lay nestled a one-month period in which the two planets were properly aligned for such a mission. Since nature offers the scientific community this bonus only once every 20 years, JPL developed a pair of twin spacecraft named Voyager, each to be launched the same month, with Voyager 2 equipped to back up Voyager 1.

At JPL, Charles Kohlhase and his team of trajectory designers and navigators were charged with the job of ensuring that his two charges would fly the best trajectory while ensuring that they could be navigated successfully within the limitations of a finite amount of propellant. The team were guardians of an open secret – Voyager 2 could continue from Saturn to Uranus and Neptune.

This remarkable opportunity was courtesy of a cycle that lines up the four planets so that gravity assists can move a spacecraft past each in a relatively short twelve-year 'Grand Tour.' It was a tempting, piquant prospect. – the last time this four-planet opportunity presented itself was in 1801, and the next would be in 2153! But NASA wanted nothing of it. The Voyagers were expected to inspect a very promising scientific target, Saturn's moon Titan. If Voyager 1

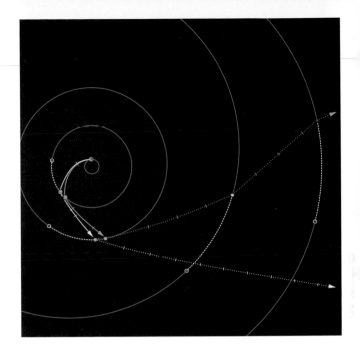

Key:
—— Voyager 1 trajectory
—— Voyager 2 trajectory

Designed to tour the outer planets, the Voyager spacecraft saved propellant by using Jupiter's gravity to slingshot themselves deeper into the solar system. After its encounter with Jupiter, Voyager 1 went on to make a Saturn flyby, while Voyager 2 used gravity assists to reach not only Saturn but also Uranus and Neptune. Both craft will continue to fly away from the Sun forever.

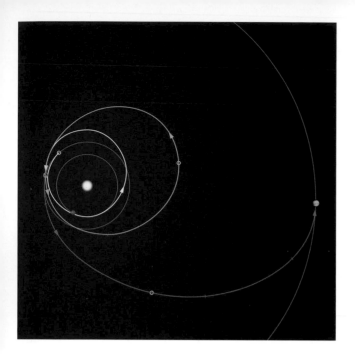

One of the most ingenious and complex uses of a gravity assisted trajectory began with the 1989 lauch of the Jupiter bound Galileo spacecraft. Taken into low Earth orbit by the Shuttle, Galileo was then propelled *inward* towards Venus, rather than outwards towards Jupiter. The first gravity assist from Venus put Galileo on an orbit that took it slightly beyond Earth's orbit. On Galileo's next pass around the Sun, a second gravity assist (this time from Earth) put the spacecraft on an orbit that took it halfway to Jupiter and allowed a flyby of two asteroids, Gaspra and Ida. Finally, on its third pass around the Sun, a gravity assist from Earth put Galileo on a direct trajectory to Jupiter.

failed in any part of this mission, Voyager 2 would have to take over, and it would have been utterly impossible to make the Grand Tour to the other gas giants. Kohlhase and crew held their tongues, and waited it out.

Kohlhase dismissed a 1976 launch because the trajectory would take his craft too close to Jupiter's hazardous radiation field, and discarded 1978 because it wouldn't permit close encounters with the Galilean moons. The target year, 1977, offered a 30-day launch period, with one-hour firing windows.' But the range of arrival dates was mind-boggling. It reached from March 1979 to September 1979 in the case of Jupiter, and from late 1980 to early 1982 in the case of Saturn. In all, there were roughly 10,000 trajectory possibilities for the two Voyagers. How, then, to sort out the best?

The solution required more than calculating which routes would take the spacecraft past the most moons as they flew by Jupiter and Saturn. The Kohlhase team had much more to consider. How much propellant would be needed? How much sunlight would be shining on the extraterrestrial 'real estate' as the spacecraft swung by? Would the communications line of sight be preserved when needed? And what about the risks, such as particle impacts from the rings, and the immense radiation field? 'We knew that an exposed human choosing to joyride the Jupiter swingby would suffer a radiation dose nearly 500 times the lethal level,' said Kohlhase, 'so that matter was a calculated risk even for a machine.'

After painstaking analysis of all of these factors and more, the team came up with 98 candidate flight plans to cover the extensive launch-date/arrival-date space, while allowing for possible launch delays. From these, after analysis of the Titan III-E/Centaur booster's capabilities, the two final trajectories were chosen. One would would skim Voyager 1 fairly close to Jupiter and take a close look at Io, then proceed to Saturn and Titan. With the other, Voyager 2 would take a more cautious route, farther from Jupiter, making additional images of the Galilean moons before continuing onward to the ringed planet.

Kohlhase took his time in asking the NASA people back in Washington to approve his plan to take Voyager 2 from Saturn to Uranus and Neptune. He always believed, with Virgil, that fortune sides with those who dare. He and his team made sure that the spacecraft's orbital dynamics around Saturn and its moons – and its fuel supply – would make it possible to visit the two outrigger gas giants. And that they did. Voyager 2's trajectory corrections were calculated by human navigators, then uplinked to the spacecraft so that it would encounter Uranus, then Neptune, with the stunning results that we will see in Chapter 9. Finally it left the solar system and joined Voyager 1 as a robotic hero of the Voyager Interstellar Mission.

In the case of Voyager 1, Jupiter's gravity redirected it to Saturn in a jolt that came from a speed increase relative to the Sun of around 16 kilometres per second (35,700 miles an hour). Remarkably, the gravity-assist manoeuvre slowed the planet by a tiny amount: in the case of Jupiter by the equivalent of about 30 cm (1 ft) per trillion years. Gravity-assist swingbys of Saturn and Uranus helped reduce the Voyager 2 trip time to Neptune by nearly 20 years, as compared with direct Earth-Neptune route.

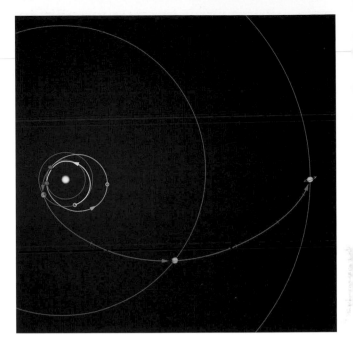

Key:
——— Final trajectory to Saturn
——— Inner solar system orbit 2
——— Inner solar system orbit 1

Cassini, NASA's latest interplanetary mission, is scheduled to arrive at Saturn in July 2004. Launched in October 1997, it used two gravity assists from Venus, and one each from Earth and Jupiter, to gain the necessary velocity to reach Saturn.

JPL scientists had to devise a special gravity assist trajectory for the Galileo spacecraft, which was blasted into interplanetary space from the Space Shuttle in 1989. For astronaut safety reasons it was equipped with a rocket system that was less energetic than the liquid-fuel Centaur originally tagged for this mission. The Centaur could have boosted Galileo on a flight path that would take it directly to Jupiter, but the replacement two-stage solid rocket system didn't have the power.

To prevent further delays to the spacecraft, mission designer Roger Diehl and his team found a new use for a computer program originally intended to help plan Galileo's orbit from one Jovian moon to another. As a result, Galileo became a 'solar cruiser,' falling initially toward the Sun to pick up a gravity boost from Venus (it had to carry high-tech parasols to protect it at this point), and then swinging back past Earth twice. At each encounter, it picked up more speed, before finally cruising off on the last leg to Jupiter.

The Cassini spacecraft, launched on October 15, 1997, for a journey of six years and nine months, was bound on an even wilder journey around the solar system. The plan was to send it, slingshot style, past Venus twice, then back past Earth, and then past Jupiter, arriving at Saturn on July 1, 2004. Saturn is nearly ten times farther from the Sun than Earth – about 1.43 billion kilometres (889 million miles). Without gravity assist, the mission would have been impossible. The craft would have had to leave Earth at 10 kilometres per second (6 miles a second), and even the mighty Titan IV with its Centaur upper stage can 'only' manage 4 kilometres per second (2.5 miles a second) with such a massive payload.

Back in July, 1965, people around the world turned on their television sets to see the first pictures from the Mariner 4 spacecraft. The pictures – sent dot-by-dot from Mars – were surprising. It was as though by some massive mistake the spacecraft was flying by the wrong planet. Many people had thought they would see a world with canals and oases! But of course they were due for a disappointment – the pictures showed a landscape of craters that looked rather similar to those of our Moon.

That day marked the end of one era and the beginning of a new one. No longer would people think of space as something completely beyond their grasp. They had seen Mars, and in a sense they had participated in planetary exploration. We experienced a touch of the scientist's thrill of discovery – overnight.

Our explorations eventually expanded from the terrestrial planets to the outer planets, in the knowledge that each of these other worlds would also have unusual features and a different physical 'personality' from the rest. We had learned a lot from observations with Earth-based telescopes. If we knew this much from just looking at their cloud cover and outer atmospheres, then it followed that a vast number of exciting new discoveries would come to light as we approached them closely and sent probes deep into their atmospheres.

When we're talking about exploration on Earth, we can resort to terms and to conditions with which we are familiar. Most people know something about the resources of Earth, the equipment needed to recover them, the rudiments of processing them, and the consumer products that will result from them. When it comes to the whys and hows of planetary exploration, however, this kind of rationale is almost completely lacking. We still know so little about the planets – and particularly the outer planets – that it's very difficult even to guess what resources they hold and whether they would be useful to us.

Then what are we looking for? Certainly not hydrogen, helium, or methane in this stage of our civilization; certainly not gold, platinum, or uranium; certainly not in the immediate future new habitats for human families. Some philosophers contend that our explorations are an end in themselves, a technological contribution to the inquisitive, adventuring nature of humankind.

Extraterrestrial life, smart and otherwise

More than three decades ago, the people of Earth sent a message to the thinking beings of the galaxy. It consisted of a series of line drawings, on a plaque attached to the Pioneer 10 and 11 spacecraft. A man and woman stand before an outline of the spacecraft, the man's hand raised in a gesture of goodwill. Hydrogen, the most common element in the universe, is illustrated in schematic form. Diagrams show the position of the Sun relative to 14 pulsars and the centre of the Milky Way, and the planets of the solar system with their distances in binary code.

SETI@home is using the computing power of the Internet to search for life outside Earth. With over three million users and one million years of CPU time they have performed over 10^{21} calculations – by far the largest computation ever performed (at least on Earth!).

So far, no response. But thousands of people are busy listening for the signs of intelligent life that might be coming *this* way, inbound to planet Earth. A number of institutions around the world devote their time and energy to the discovery of sentient beings elsewhere in the universe. Foremost among these is the U.S.-based Search for Extraterrestrial Intelligence (SETI) Institute, which uses the vast 305-m (1,015 ft) Arecibo dish in Puerto Rico together with a smaller telescope at Jodrell Bank, England, for much of its work.

Astronomer Jill Tarter is director of SETI's Project Phoenix, which monitors 1,000 nearby stars in the hope of finding radio signals from intelligent beings. 'If we detect a signal – even a cosmic dial tone with no information content – we will have answered a question that is at least as old as recorded history: 'Are we alone?'' she says. 'We will know that intelligent life is a common end product of the laws of physics and chemistry that govern the universe.

'In the long run, I think this would be as significant as Darwin's theory or Copernicus's heliocentric model of the solar system. It helps us see ourselves in a bigger picture. Also, the differences between humans and another species that has evolved somewhere else will trivialize the differences among humans that we find so difficult today.'

One of the mainstays of the SETI enterprise is SETI@home, in which enthusiasts with internet-connected computers download and analyse radio telescope data. The project claims an amazing three million users, and by 2002 had logged over one million years of computing time.

The cause of astrobiology – searching for and studying any kind of extraterrestrial life, is served by numerous organizations around the world. These range from the SETI Institute to the large astrobiology group at NASA Ames Research Laboratory in California, and the Centre for Astrobiology in Wales, which has a special interest in the theory of panspermia – the idea that biological molecules and cells are present in comets and interstellar dust.

Interplanetary Voyagers

'Through Voyager the human intellect has extended its horizons to the farthest reaches of the Solar System, initiating a new Age of Discovery, and giving mankind a deeper insight into the role of intelligence in the Universe.'

Thomas Paine, NASA

As we extend our explorations of the solar system, the dream is to see the previously unseen, and to understand more of the mysteries of our universe. Whether this dream continues to come true depends largely on spacecraft – the robots and Earth-orbiting eyes in the sky typified by the Hubble Space Telescope. There now exist whole generations of watching, listening, moving, remembering, communicating space creatures, made by human hand, and served by earthbound computers that sift, sort, and display their findings.

If some sentient extraterrestrial being were to make contact with an Earthling spacecraft, its first reaction would almost certainly be to identify this high-tech contraption as an intelligent visitor from inner space. As astronomer J. Allen Hynek said back in 1981, 'It's our UFO to somewhere else!'

In fact, like a living thing, every spacecraft must be endowed with senses and memory, the ability to communicate, and power to move from place to place and run its internal systems. Behind the scenes, however, are the innovators and the puppet-masters. The largely unsung heroes of space exploration are the spacecraft engineers. These are the people who must build a craft that will work without fail on these incredible journeys into space. While the first decades of the Space Age produced volumes of fresh knowledge, new missions require much more than merely modifying and scaling up already-proven ideas. Each new space programme poses a new set of human challenges. It's a huge job, for these craft are not only intricate but also large – at launch Cassini weighed about the same as an empty 30-passenger bus.

Like any other engineers, these men and women work with electronic blueprints and deal with contractors – but they must also predict how well the craft will perform in the peculiar conditions of heat, radiation, and electric and magnetic fields. They must be sure that there is enough power available to operate the instruments, and that these instruments can be pointed in precisely the right direction when needed, and held steadily enough to make immaculately crisp images. And they must guarantee that radio antennae can be pointed back toward Earth to send precious data and images home.

While some leeway is available for overall spacecraft design, a few general guidelines are unchangeable. The spacecraft must be compatible with the launch vehicle, for example, and also with the capabilities of Earth stations – for example NASA's Deep Space Network (DSN) or ESA's ESTRACK Network. It must also carry its own power supply – solar cells and/or generators powered by the heat of radioactive decay.

Other essentials include:

- a system to sample instrument readings and radio the information to Earth.
- a system to control on-board equipment both automatically and by ground command;
- a design that guarantees instruments will be protected from vibration and magnetic and electromagnetic interference; and
- maximum utilization of earlier experience in design, development, construction, and testing – as well as in reliability and quality assurance programmes.

While American and Soviet spacecraft were visiting the terrestrial planets in the mid-1960s, planners at the Illinois Institute of Technology investigated general needs for an outer-planets mission, and JPL began to map the details of an actual spacecraft for a 'Grand Tour' beyond the asteroid belt.

JPL's early planning had to tackle three main questions. Could scientific instruments be designed to perform reliably throughout a nine-to-twelve-year space flight? Could engineers make the spacecraft self-reliant enough to allow quick reactions in emergencies? And could the hazards of the asteroid belt (later found to be quite insignificant) and Jupiter's radiation environment be predicted well enough to ensure that a truly hardy spacecraft would be built?

Prototypes of new components were designed and produced; ways in which the systems might interact were analysed; methods of ensuring reliability and quality were studied. Then, in 1970, JPL announced TOPS – the Thermoelectric Outer Planets Spacecraft concept.

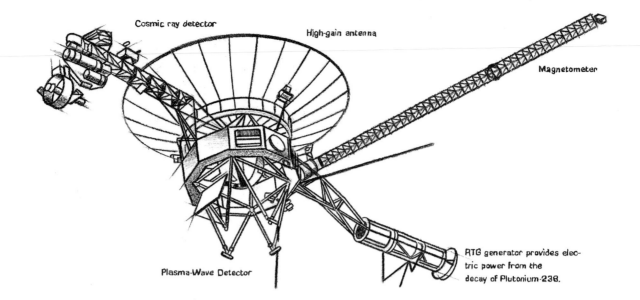

Cosmic ray detector

High-gain antenna

Magnetometer

RTG generator provides electric power from the decay of Plutonium-238.

Plasma-Wave Detector

TOPS was never built, but it was a pathfinder nevertheless. We can look back upon it as a generic outer-planets design, containing many of the elements that would eventually be utilized in the Pioneer, Voyager, Galileo, and Cassini spacecraft.

One feature set this spacecraft's external appearance quite distinctly apart from all the earlier planetary explorers – a twelve-foot antenna dish that dwarfed all the other systems and made the spacecraft look rather like a giant camera flashgun.

A large box-like structure (the bus) was connected rigidly to the antenna. This housed electronic and propulsion equipment and protected it from radiation and micrometeoroid hazards (eventually the Voyager spacecraft would carry 22 kg [48 lb] of tantalum shielding for this purpose). Sprouting from the bus were small thruster nozzles for correcting the trajectory and attitude of the spacecraft – plus a smaller antenna and two 9-metre (30-ft) booms on which experiments could be mounted outside the diameter of the large parabolic dish.

Aside from the equipment compartment and the large antenna, the largest structures were a special platform for experiments that needed to view targets directly, and radioisotope thermoelectric generators to provide the spacecraft's entire power supply. The outer planets mission could not be solar-powered (at Neptune, for example, solar radiation is only 0.1 per cent as strong as it is at Earth); thus RTGs were utilized, as they were for the Nimbus and Transit Earth satellites and earlier planetary probes. To power the Galileo spacecraft with the Sun's energy would have required use of 182 square metres (2,000 square feet)

Voyagers 1 and 2 were NASA's most successful interplanetary missions. Most of the information we possess on the outer planets came from these two spacecraft. Voyager 1, launched second but placed on a faster trajectory, flew-by Jupiter in March 1979 before reaching Saturn in November 1980. Voyager 2 reached Jupiter in July 1979, Saturn in August 1981, Uranus in January 1986 and Neptune in September 1989. Travelling at over 17 kilometres per second (38,000 miles per hour), Voyager 1 is currently the most distant man-made object in the universe. Voyager 2 is not far behind and both craft are now exploring the region in space where the Sun's influence ends and the dark recesses of interstellar space begin. The Voyagers have enough electrical power to operate at least until 2020. By then, Voyager 1 will be 20 billion kilometres (12.4 billion miles) from the Sun and Voyager 2 will be 16.9 billion kilometres (10.5 billion miles) away.

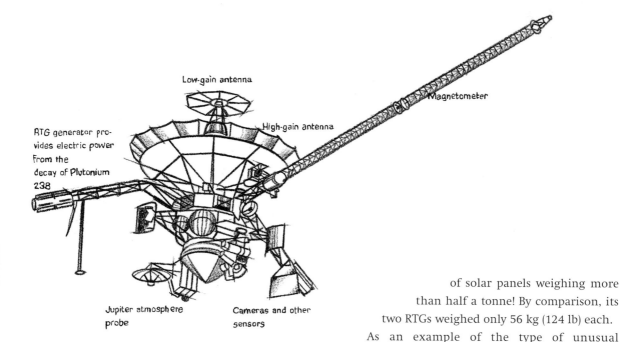

Low-gain antenna

High-gain antenna

Magnetometer

RTG generator pro-
vides electric power
from the
decay of Plutonium
238

Jupiter atmosphere
probe

Cameras and other
sensors

of solar panels weighing more than half a tonne! By comparison, its two RTGs weighed only 56 kg (124 lb) each.

As an example of the type of unusual engineering challenges tackled by people who design outer-planets spacecraft, let's look at thermal control. This is a critical subject since delicate components will age quickly with exposure to high temperatures, and at very low temperatures many electronic circuits will simply stop working.

In Pioneers 10 and 11 – the first spacecraft to venture past the orbit of Mars – thermal controls kept instrument temperatures between 32 and -29°C (90 and -20°F). Parts of their outer walls were painted black to collect the greatest possible amount of solar radiation; the instrument compartment was insulated to prevent unwanted heat loss; and louvres – actuated by bimetallic bands like a common thermostat – were provided to cast off heat when the level became too high. Later outer-planets spacecraft were equipped with radioisotope heater units about the size of a fingertip. Cassini and the Huygens probe are equipped with 117 of these little gems.

At first glance the Galileo spacecraft, built to make a close-up inspection of the Jovian system over a number of years, resembled Voyager. But big differences showed up as one looked closer. A great deal of extra equipment was attached to the spacecraft right next to the big high-gain antenna. One part of this was the retropropulsion module, which would slow the craft into orbit. The other part was the most dramatic – the atmospheric probe designed to be released from the spacecraft (several months before Jupiter orbit insertion) and penetrate Jupiter's clouds for the first time ever. The probe was

Launched from the Shuttle Atlantis in 1989, Galileo arrived at Jupiter in 1995. The mission narrowly avoided disaster after the high-gain antenna failed to deploy. However, new data-compression routines allowed the low-gain antenna to transmit information back to earth – albeit at a slower rate. Upon arriving at Jupiter, Galileo released a probe that sliced into Jupiter's atmosphere. Descending to a depth of 150 kilometres (95 miles) the probe collected 58 minutes of data on the local weather before succumbing to the intense heat of the atmosphere. NASA has repeatedly extended Galileo's original two-year mission, but knowing they will eventually lose contact with the spacecraft, they have chosen to plunge Galileo into Jupiter's atmosphere in September 2003. This is to ensure that it will not hit Europa, where some believe there may be extraterrestrial life.

equipped with parachutes to ease it into the Jovian atmosphere – much as the Soviets sent their Venera 4 probe through the clouds of Venus in 1967. Its mission was to sample Jupiter's upper layers with half a dozen scientific instruments and radio its findings back to the orbiter part of Galileo as it approached Jupiter.

That brings us to the largest interplanetary spacecraft ever built, Cassini, two storeys high and four metres (13 ft) wide, and weighing 5.6 tonnes at launch. The spacecraft is topped by a four-metre-wide antenna; its 'feet' are three RTGs, arranged around two rocket engines. Reaching 11 metres (36 ft) out to the side, like a long thin arm, is a magnetometer boom.

From antennae to rocket engines, the spacecraft represents a 'stack' of rugged engineering. Three major structures are attached to the stack. One is the Huygens probe, an ESA-designed spacecraft for exploring the atmosphere of Saturn's moon Titan. The other two are pallets, individual platforms that support scientific equipment – one set for remote sensing, the other for measuring fields and particles. A mapping radar and rigidly mounted sensors are attached to the top end of the stack, which also houses the orbiter's electronics systems.

The big dish is a high-gain antenna, a highly directional system capable of amplifying signals at great distances; perched on top of it is one of two low-gain antennae (offering less amplification but requiring less accurate positioning). The other low-gain antenna is mounted near the bottom of the spacecraft. Both were designed to transmit data and receive commands. A low-gain antenna can be used when, for whatever reason, the large dish cannot be pointed at Earth.

The Cassini orbiter carries equipment for twelve long-term experiments, and the Huygens probe is equipped to handle six relatively brief experiments. That's a lot of experiments, requiring 18 sets of scientific instruments capable of running for a number of years, and it took a lot of smart engineering to make this possible.

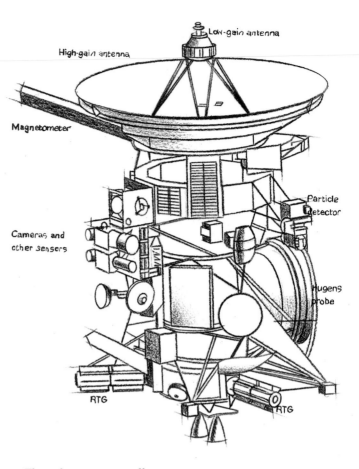

One of the largest (6.8 metres [22.3 feet] high), heaviest (5,600 kilograms [12,346 pounds] at launch) and most complex interplanetary spacecraft ever built, Cassini will survey Saturn's weather systems, atmosphere, rings, magnetic field and extensive moon system. Hitched to the side of NASA's Cassini is the ESA's Huygens probe. Huygens has been designed to plunge into the atmosphere of Titan, the only known moon in the solar system with a sizeable atmosphere. Cassini was launched on 15 October 1997 and is expected to arrive at Saturn on 1 July 2004.

Let's take a look inside Cassini's brain, a neat electronic package called the command and data subsystem (CDS), coupled to a compact, fault-resistant memory with no moving parts, the solid state recorder subsystem (SSR).

The CDS performs many vital jobs. It picks up commands from Earth, processes them, and makes sure that other subsystems carry them out. It is the place where spacecraft data are gathered, converted into a useful format, and sent to Earth. It keeps time, controls spacecraft temperature, and keeps a lookout for equipment failures. If there is a failure, it's up to the CDS fault protection software either to correct it or to keep the spacecraft safe until controllers back on Earth can diagnose the problem and send it instructions on how to repair it or work around it.

Of the many parts of the CDS, the most important is an IBM engineering flight computer (EFC), selected for Cassini because of its previous success in avionics (aviation electronics) applications where extremely high reliability is required. It includes features that protect against corrupted information or computer 'crashes.' Taken in a broader sense, its capabilities increase the lifetime and overall fault tolerance of the spacecraft's electronic systems.

The computer's central processing unit operates at less than one per cent of the speed of ordinary desktop computers, but unlike them it is robust and radiation-resistant. In the late 1990s, NASA engineers were working to increase performance to more than 200 MHz (from 20 million to 200 million operations per second) without losing reliability. This will permit future spacecraft to operate more independently and to do much more on-board data processing.

The CDS 'brain' is connected to the spacecraft's sensory system by a network of electronic nerves called the bus interface system. The BIS fans out to remote terminals embedded in the many individual subsystems that perform commands and data gathering. The terminals collect science data from the instruments and engineering information such as temperatures and hardware status from all over the spacecraft, and bundle it into data 'packets' for BIS delivery to the central computer.

Every brain must have a memory, and Cassini's is no exception. Onboard recording devices are needed to store information about science experiments and the spacecraft's state of health, and then to play this information back at a later time. During close encounters with planets and their moons, spacecraft rarely relay all their experimental data in real time to the DSN or ESTRACK. Instead they store most of it for later playback. While previous missions used flight tape recorders, Cassini engineers decided to use solid state recorder (SSR) technology for bulk data storage and playback. On the whole, recorders gave good service, but they are relatively bulky, their moving parts can wear out, and the magnetic tape is coated with oxide particles that wear off with time. Two SSRs made it possible to avoid all of these potential problem spots while providing 2 billion bits (2 Gigabits) of storage capacity each. Both SSRs have enough spare memory to ensure that they still have at least 1.8 Gigabits of storage at the end of the 12-year mission. Finally, the SSRs are smart. Each can store critical flight programmes, keep track of its temperature, and monitor its internal energy level.

And there you have it – Cassini/Huygens, the most advanced robotic creature in the evolution of spacecraft. The next generation will likely be small, swift, specialized craft. And then, perhaps, we will design a craft to send people to the planets, beginning with Mars.

Why do we do it? Again, despite the interest in extraterrestrial natural resources, and despite the clues that we think we may find out there to explain our own planet and its dynamics, the greatest and perhaps the best reason for space exploration is to serve our sense of wonder. And the craft itself is, in the words of Norman Ness, a specialist in planetary magnetic fields, 'a unique artefact with which to extend the sense of mankind in our universal search for truth and understanding.'

Robot Scientists

Many decades ago, even before the first jet plane was flown, the great Russian space pioneer Konstantin Eduardovich Tsiolkovsky (1857–1935) – embittered by the lack of recognition he received for his work, eternally embarrassed by his own deafness, and haunted by the promise of space for humankind – took pen to paper to rebuff his critics.

'There was a time [he wrote], not long ago, when the idea that it might be possible to know the makeup of the heavenly bodies seemed ridiculous, even to eminent scientists and thinkers. That time is now past. I think that the notion of a closer, direct study of the universe will seem even stranger at the present day . . . [but] the first use of jet-craft will open a new era in astronomy – an era of more intent study of the heavens.'

Tsiolkovsky died two decades before the advent of Sputnik. His ideas met with some scepticism; yet thirteen years after his death, when Jesse L. Greenstein of the Mount Wilson-Palomar Observatories told the University of Chicago's Symposium on Planetary Atmospheres that 'at present there is a definite possibility of establishing a 'satellite-rocket' observing station rotating about the Earth at 500-kilometres altitude,' nobody laughed. Neither did they laugh when Greenstein went on to report that 'present fuels even permit escape of a multistage rocket from the gravitational field of the Earth.'

Rocket-borne scientific measurements of the Earth's upper atmosphere began in 1946 with several programmes in which Greenstein was only one of many participants. The first hint of what lay beyond the Earth's immediate environs, however, came in February, 1958, when an instrument hovering in near space exhibited some very peculiar behaviour.

On this occasion Dr. James A. Van Allen had provided a Geiger-Müller (GM) counter for Explorer I, which was to become the United States' first successful satellite. Scientists had used the GM tube (also known simply as the Geiger counter) for some time to detect levels of radiation. It contains a positively charged wire separated from its negatively charged container walls by a gas-filled space. The tube produces an electrical pulse when it is penetrated by a charged particle such as an electron, so that the number of counts per minute indicates the quantity of particles and radiation present.

After Explorer I's launch, the jubilant scientists hurried to their tracking stations to listen to the first American telemetry signals from orbit. Disappointingly, Van Allen's Geiger counter stopped working when the satellite reached the highest point (apogee) of its trajectory, so it was believed something had gone wrong either with the counter or with the telemetry system. Then, miraculously, when the satellite approached its low point (perigee), the signals started again.

There could be only one reasonable explanation for this odd behaviour – radiation levels had been so great that the counter had become saturated with more particles than it was believed could exist in that region of Earth's environment. The now-famous Van Allen Belts had been discovered – belts formed by Earth's magnetic field, which expand and flatten as they are 'blown' back by particles from the solar wind, and which are believed generally similar to the radiation belts detected at Jupiter.

Since that historic discovery, a huge number of instruments have been designed to act as our eyes and ears in space. Within the next decade and a half, satellite-borne instruments were to examine the Sun and universe in regions of the spectrum that we cannot see from Earth; planetary spacecraft were to visit Mars and beam back incredible photographs of its surface; the United States' Surveyor craft would scrape up samples of lunar soil and analyse them, and a Soviet spacecraft would perform a similar landing, picking up 'Moon rocks' and flying them back to Russia.

EXPLORER 1
AMERICA'S FIRST EARTH SATELLITE

JET PROPULSION LABORATORY

CALIFORNIA INSTITUTE OF TECHNOLOGY

Space Propulsion: The Early Days

Above: Konstantin Tsiolkovsky's 1903 design for a hydrogen-oxygen powered spaceship. Tsiolkovsky said 'Earth is the cradle of the mind, but one cannot live in the cradle forever'.

Right: Five rocketry pioneers pose with scale models of the missiles they created in the 1950s. From left to right: Dr. Ernst Stuhlinger; Major General Holger Toftoy; Professor Herman Oberth; Dr. Wernher von Braun; and Dr. Robert Lusser. This photograph was taken on 1 February 1956, two years before the launch of Explorer 1.

All chemical rocket systems use the familiar principle of internal combustion. Unlike cars and trucks, however, rockets must carry their own oxidizers, because air is only briefly available and indeed may not be the preferred oxidant. The fuel and oxidant – referred to as propellants – ignite and the resulting gases expand with heat. As they are forced through a funnel-shaped nozzle, the engine obeys Newton's Third Law. Since every action has an equal, opposite reaction, the gases pushing one way force the vehicle in the opposite direction.

The staged rocket, featured today in all large launch systems, was detailed in 1650 by Kazimierz Siemienowicz (c.1600–c.1651) of Poland, who illustrated his book *Artis Magnae Artilleriae* with drawings of three-stage rockets, clustered rockets, and other ideas that were centuries ahead of his time.

Siemienowicz and the rocketeers of his day had nothing but gunpowder for their fuel, and their goal was nothing more ambitious than to build a better bomb-carrier. But in 1881 Nikolai Ivanovich Kibalchich (1854–1881) of Russia designed a rocket aeroplane – while he was imprisoned for participating in an attempted assassination of Tsar Alexander II.

By 1898, Konstantin Tsiolkovsky (1857–1935) had documented the basic principles of rocket motion. At the time the Wright Brothers were pioneering aviation in the U.S., he was already writing about satellites, space suits, and regenerative systems for long-term human space flight.

Tsiolkovsky had predicted that liquid propellants would power more efficient rockets than solid propellants. Then Robert H. Goddard (1882–1945) of the United States showed it was true. Goddard worked on solid rockets during World War I, but these could not reach the altitudes he wanted for weather research. He took the first giant step toward realizing his dreams on March 16, 1926, when he fired a liquid oxygen-gasoline rocket to an altitude of 55 metres (184 ft) near Roswell, New Mexico.

Three years before Goddard's historic rocket shot, Romanian-born Hermann Oberth of Germany (1894–1989) published a detailed booklet that prophetically described large interplanetary space vehicles, space stations, and ideas for ion and electric propulsion. Beginning in the 1940s, Willy Ley (1906–1969) and Arthur C. Clarke (1917–) were among the many scientist-writers who built the case for spaceflight, along with the greatest science popularizer of them all, Isaac Asimov (1920–1992).

Tsiolkovsky knew all about this many generations ago. 'Earth is the cradle of the mind,' he philosophized, 'but one cannot live in the cradle forever.'

While instruments cannot possibly command the judgement and discrimination of an astronaut, their 'senses' encompass the whole electromagnetic spectrum. They require no life-supporting equipment, and, in most cases, they simply don't need an astronaut to turn them on and off. Though some people will see less glamour in sending a few hundred kilos of instruments into space, most will agree that these programmes are, relatively speaking, much more practical. And in terms of missions to the outer planets, we must admit that space scientists would have had to delay their exploratory plans by many decades, had we insisted upon sending humans.

Science and Space

As might be expected, there's a huge competition for space on any spacecraft, for many scientists have promising experiments that they want to carry out. Finally, after all the proposals have been reviewed and ruled upon, a cadre of principal investigators remains. In the case of Cassini, 27 different science investigations swelled the overall team – including those from industrial organizations – to more than 4,300 dedicated men and women. They include scientists, mission designers, spacecraft engineers, and navigators.

Long before scientists were invited to submit their ideas, mission analysts had been wrestling with the question of what might make up the ideal feasible payload. 'Since it was not possible at that time to specify the science instrumentation that would definitely be selected for these missions,' said JPL's James E. Long, 'a broad range of instrument capability was considered so that detailed spacecraft design studies could be made.

'It was concluded that a balanced science capability of fields-and-particles instrumentation and imaging instruments requiring precise pointing at the planet during closest approach would be required to satisfy the major interplanetary and planetary exploration objectives.'

Though this was to be an entirely new spacecraft with completely new goals, its planners could turn to experiments used earlier to investigate planets closer to the Earth. Among them:

- Imaging equipment that studies planetary atmospheres – their circulation, colour, cloud structure, and internal makeup – along with features like satellite surfaces, planetary rings, and lightning and auroral activity. Typically these use Charge Coupled Devices (CCDs) for highest performance and resolution.

- Photometers that detect ultraviolet radiation in space and in planetary atmospheres, revealing information on helium-to-hydrogen ratios and charting the structure of planetary surfaces, satellites, and rings.
- Radiometers and spectrometers that map planetary temperatures and atmospheric movements, and the composition of the atmosphere and surface, through detection of infrared light emission and absorption.
- Radio receivers that detect radio energy produced by the planet or its charged-particle environment.
- Radio equipment that (though used mostly for communication) sometimes provides additional information by measuring gravity fields and revealing how the planet's atmospheric gases, magnetic fields, and rings, affect the radio signal.
- Magnetometers that reveal the magnetic characteristics of the planets and the interplanetary medium. (The nature of a magnetic field gives clues as to whether a planet has a metallic core, like Earth, and/or some other conductive material such as salt water.)
- Micrometeor detectors that chart the size and velocity of particles encountered in the asteroid belt and beyond.

All these devices – and variations – were included in the 'baseline science payload' for the first outer-planets mission. The same techniques had been used elsewhere in the solar system; but never before had they been called upon to operate for such a long time in space. With the addition of spectrometers, instruments like these proved to be mainstays of future outer-planets missions. The all-important spectrometers would determine what molecules were present in planetary atmospheres; plasma and particle detectors would detect and measure ions and electrons.

Special missions naturally call for special experiments. The Galileo atmospheric probe instruments included devices for detecting clouds, lightning, and the presence of helium. Among Voyager's science instrumentation was IRIS (Infrared Interferometer Spectrometer and Radiometer), which determined temperatures, measured reflected sunlight, and detected the presence of certain chemicals. In the gas giants it not only found hydrogen and helium, but also water, methane, acetylene, ethane, ammonia, and phosphorus and germanium compounds.

Planetary Probes

To pin down the composition and dynamics of the planets with any great accuracy requires sending instrumented probes into their atmospheres – as with the Galileo and Cassini spacecraft. A mother spacecraft flies by or lingers in orbit around the planet, dropping a probe containing instruments for measuring magnetic fields (magnetometers), chemical species (spectrometers), atmospheric vapour (nephelometers), and other properties. As it falls, and as long as it can withstand heat, pressure, and radiation, the probe radios its findings back to the mother ship for relay to Earth.

These probes are extremely important. The limitation of fly-by missions and orbiters is that methods like ultraviolet spectrometry collect information mostly about the upper atmosphere, since light of these wavelengths from lower levels is absorbed and cannot be detected. This means that instruments tend to register only the lighter elements present in a planetary atmosphere. The heavier elements are held nearer the planet's surface by gravity.

A spectrometer looking at the upper layer of Earth's atmosphere, for example, would reveal a large amount of hydrogen, for, even on our own planet, gravitational separation begins to occur noticeably at an altitude of about 80 kilometres (50 miles). If limited to this type of measurement, a visiting 'flyby' spacecraft would give a very misleading picture of an atmosphere that is almost 99 per cent nitrogen and oxygen at sea level.

Power for Space

Spacecraft are generally powered by solar cells and/or 'nuclear batteries' – Radioisotope Thermoelectric Generators (RTGs). If the mission stays relatively close to the Sun, for example in and around the terrestrial planets, the amount of solar radiation is generally sufficient to allow solar cells to run the system. Beyond the asteroid belt, however, the solar flux thins and RTGs must be used.

Solar cells, otherwise known as photovoltaic (PV) cells, perform the technical miracle of using photons of light to make electricity run through a circuit. This is done by exposing a sandwich of two dissimilar semiconductors to light. In a conventional silicon PV cell, the transparently thin layer of the sandwich facing the sun contains a trace of boron (giving it a predominantly positive atomic charge); the other side is 'doped' with a trace of phosphorus (giving it a predominantly negative atomic charge). When photons arrive at the junction between the two layers, the normal pairing of electrons and their

The Seebeck Effect

To trace the history of thermoelectricity, we must go back to the year 1821 and to the Berlin laboratory of Thomas Johann Seebeck (1770–1831). In that year, Seebeck noticed that an unusual effect occurred if he connected certain materials together and then heated the point at which he joined them. He mistakenly thought he had produced a magnetic field from the temperature difference. We know today that his experiment had caused a buildup of electrical potential (electromotive force) at the unheated ends, where it can be easily measured. It is still called the Seebeck Effect.

Seebeck did not develop a practical application for his discovery, but the common metallic thermocouple – which works on the same principle – became one of the first 'direct energy conversion devices' when it was put to practical use for measuring temperature.

It's inefficient to use metals to produce electricity with the Seebeck Effect. For this reason, nothing very exciting happened in the field of thermoelectricity until the development of semiconductors – materials like germanium, silicon, gallium-arsenide, lead-telluride, and indium-antimonide. Normally, semiconductors are more resistive to conducting electrical current than metallic conductors. The secret is that they can be given predominantly negative or positive charges by adding other chemicals, a process known as 'doping.'

The semiconductor used in present-day RTG thermoelectrics is a compound of silicon and germanium (SiGe), which has a conversion efficiency rated at 7 per cent. (Conversion efficiencies measure efficiency in transforming thermal energy to electrical energy. Hence a 7 per cent efficiency SiGe thermocouple should produce 14 watts of electricity from a heat source radiating 200 watts of thermal energy.)

The dissimilar materials used in the SiGe thermocouple are phosphorus and boron. One part of the couple is 'doped' with a trace of phosphorus. Each atom fits neatly into the

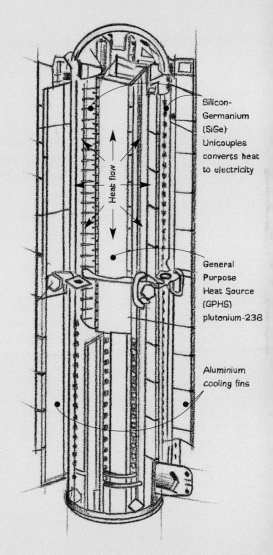

Heat flow

Silicon-Germanium (SiGe) Unicouples converts heat to electricity

General Purpose Heat Source (GPHS) plutonium-238

Aluminium cooling fins

Radioisotope Thermoelectric Generators (RTGs) are the only practical source of power for a spacecraft operating beyond the orbit of Mars. Because they contain no moving parts, they also have the advantage of high reliability.

semiconductor's structure, but each atom of phosphorus has an 'extra' electron compared to its neighbours, so the doped material ends up with an excess of negative charge. The boron used on the other side of the couple, meanwhile, contains one less electron than its surroundings, resulting in a predominantly positive charge). Supercharging the couples with negatively charged electrons and positively charged 'holes' makes them more efficient thermocouples than metals.

Now imagine cylinders of these two materials placed side by side on a hot plate. The hot plate is perhaps 370°C (700°F); the air around the other ends of the cylinders is cooled with a fan. If you connect the two cool ends of the cylinders with a wire or some other conductor, current will pass through it as electrons are forced from the phosphorus-doped (negative-type) cylinder through the conductor and into the boron-doped (positive-type) cylinder. The beauty of this system is that the current can be made available to perform work without need of a single moving part.

The efficiency of the device increases as the difference in temperature – the temperature gradient – between the hot and cold junctions increases. Thus thermoelectric devices work better in the extreme cold of space than they do in the relatively warm confines of Earth's atmosphere.

If the hero of thermoelectrics was the semiconductor, then the key to its greatest success – thus far at least – is the radioactive isotope. Isotopes are varieties of atoms of a given element; radioactive isotopes are unstable types that decay to more stable elements by giving off some of their energy, which can be used to heat the thermocouple.

Some 1,300 different species of radioisotopes are available as fuels; however, fewer than a dozen may remain after those with unfavourable characteristics have been culled from the list. It may be wise, for instance, to eliminate those candidate fuels with half-lives of less than 100 days or more than 100 years; those with relatively poor thermal properties; those which give off dangerous gamma-ray emissions; and those which are extremely expensive.

counterpart holes is disturbed, and the two are freed to supply energy to an external load. In a spacecraft, this load might be the power-conditioning unit for a science experiment, or a storage battery. (See the box on p.103 for a fuller explanation of semiconductors.)

Solar cells have been used in spacecraft since the late 1950s, supplying power to some early U.S. Vanguard satellites and to Britain's Ariel 1 ionosphere probe in 1962. Today they are found everywhere, from pocket calculators to power supplies for outdoor lighting. Someday huge fields of solar cells will be installed on the Moon, gathering terawatts of energy that will be delivered to our electrical power grids by invisible microwave beams.

The International Space Station was designed with the most powerful solar power plant in space, dominated by four pairs of giant gold-coloured solar wings. Each of these two-wing arrays extends 72 metres (240 ft); the wings carry 32,400 solar cells and provide about 75 kilowatts of power.

Fortunately, the ISS is relatively close to the Sun, where energy is easy to come by. Things get more complicated as projects become larger and farther from the Sun. A 10-megawatt solar array would have to measure an impracticable 68,000 square metres (750,000 sq. ft) at Mars, and an unthinkable 760,000 square metres (8.4 million sq. ft) at Jupiter!

The RTG came to the rescue. The first one to come out of the lab rested on the White House desk of President Dwight D. Eisenhower on the afternoon of January 16, 1956, a peculiar-looking object the size and shape of a grapefruit.

The device weighed only 1.8 kg (4 lb) and provided a continuous power supply of only 2.5 watts (most household light bulbs consume at least 60 watts). Its remarkable feature was that it had unique staying power. It was capable of delivering 11,600 watt-hours in around 280 days – about the same energy delivered by common nickel-cadmium (NiCd) batteries weighing 700 pounds!

Just thirty months later, a Transit satellite was put into Earth orbit from Cape Kennedy, carrying the first RTG ever to operate in space. The device functioned well beyond its estimated five-year operating lifetime.

From the beginning of outer-planets spacecraft planning, it was fairly clear that nuclear power supplies would be needed to provide electricity. The duration of the mission would be much too long to consider using ordinary batteries; the spacecraft would travel too far from the Sun to use solar power; and an RTG would be more efficient than the yet-unproven space reactor concepts. Thus the original Grand Tour spacecraft concept was called the Thermoelectric Outer Planets Spacecraft because power would be supplied thermoelectrically from the heat emitted by radioisotopes.

The workhorse isotope in today's RTGs is plutonium-238 (Pu-238) – not to be confused with Pu-239, the by-product of uranium fission. It decays by giving off alpha particles, and releases heat in the process; half of the atoms in any sample will decay in 89.6 years – the isotope's half-life.

Since Pu-238 is an 'alpha emitter,' it is relatively safe. Alpha particles are hazardous only if swallowed, inhaled, or absorbed into a wound, and do not have enough intrinsic energy to penetrate this piece of paper.

At the heart of today's RTGs is an array of 'modules,' each of which contains four plutonium dioxide fuel elements, each about the size of the tip of a human thumb and coated in silvery iridium metal. The modules are surrounded by several layers of shielding materials for insulation and protection from rupturing or corrosion. Added to this are the all-important array of power-generating thermocouples wired in parallel, their hot junctions placed snugly in proximity with the plutonium capsule, their cold junctions exposed to a 'heat sink' that taps off excess thermal energy.

RTGs are the only major power supply aboard an outer-planets spacecraft (short-lived chemical batteries may be preferable for brief experiments like those on an expendable probe), and are required to run scientific experiments, communications, and other on-board systems. To meet these requirements, the first outer-planets spacecraft designers called for a cluster of four RTGs, intended to produce a minimum of 439 watts during planetary encounters. Voyager eventually carried three 150-watt RTGs. Galileo was fitted with two 296-watt RTGs. Cassini carried three RTGs identical to those on Galileo. Dubbed unimaginatively 'General Purpose Heat Sources,' they weigh about 56 kg (124 lb) each, including 11 kg (24 lb) of plutonium dioxide fuel, the thermoelectric converters, cooling tubes, insulation, and an aluminium shell. Very roughly the size of a golf bag, they are 106 cm (44 in) long and about 41 cm (17 in) in diameter. The Cassini RTGs provided 888 watts of electrical power from 13,182 watts of heat at launch; by the end of the mission, its power output was expected to drop to 628 watts.

So long-lived are RTG power sources that JPL was able to receive science data from Pioneer 10 in 2002, 30 years after the spacecraft's launch. The craft was then nearly 12 billion kilometres (7.5 billion miles) from Earth, so far away that the radio message took 22.1 hours to make the round trip at the speed of light.

Where would our exploration of the outer-planets be without RTGs? Almost certainly still on the drawing board. We would have seen none of those wondrous images of the Jovian moons and Saturn's satellites, and the idea of a robot expedition to Pluto would be no more than a futuristic dream.

Rocketing into Space: The Launch System

Engineers, scientists and technicians must spend years planning the vital period from lift-off until the final insertion into interplanetary orbit.

The star of this momentous half hour will be the launch vehicle – far more than a mere rocket, for it is an entire system containing rocket engines plus its own built-in attitude and guidance control equipment. The complete Apollo/Saturn V system, which took the first humans to the Moon in 1969, contained about 100 separate rocket motors.

While Saturn V was the most powerful American launch system ever built, the most successful must be the Titan series, offspring of a Cold War missile system. The Titan IIIE/Centaur, standing 16 storeys high, was first used in 1974 to launch two Viking spacecraft to Mars. The Titan itself was a two-stage liquid-fuelled rocket. In this configuration it had two large solid rockets ('strap-ons') attached, one on each side. The somewhat wider, liquid-fuelled Centaur upper stage sat atop this structure, giving it a slight bottle shape, and on top of that stood the aerodynamic fairing with the spacecraft tucked safely inside. This system was used to launch the eminently successful Voyager twins in 1977. More than two decades later, a similar but more powerful Titan would send the Cassini spacecraft on its way to Saturn.

At 5.6 tonnes, the Cassini payload was more than twice the weight of Galileo. To lift it out of Earth's atmosphere, planners selected the largest expendable launch vehicle in the U.S. space arsenal, the Lockheed-Martin Titan IVB, topped with its Centaur upper stage. This would provide sufficient initial energy to send the craft on its complicated Venus-Venus-Earth-Jupiter gravity assist trajectory.

The Titan IVB-Centaur looked very much like the system used to launch the two Voyagers. With the Titan flanked by its two SRBs, the lower stage assembly resembled three giant cigars attached side by side, 12 storeys tall. Clamped atop the core section was a bulbous aluminium fairing, 4.8 metres (16 ft) across and 20 metres (66 ft) high, that contained the liquid-fuelled Centaur upper stage, the guidance and navigation system, and the reason for it all, the Cassini spacecraft. Together, the propulsion systems weighed two million pounds (1,000 tons), of which nearly 90 per cent was fuel.

The total stack height of the Cassini launch vehicle was 56 metres (184 ft), approximately the height of an twenty-storey building. Together, the launch vehicle and payload weighed in at 1,038 tonnes.

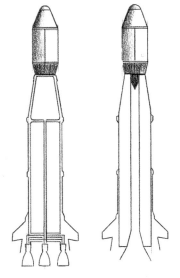

The basic principles of the rocket are simple, as these diagrams of a liquid-fueled rocket (left) and a solid-fueled rocket (right) show.

In liquid-fueled rocket engines, a fuel and an oxidizer are pumped into a combustion chamber where they are burned to create a high-pressure and high-velocity stream of hot gases. The fuel can be hydrogen (Shuttle's main engines), kerosene (Saturn V) or even alcohol (V2). These gases are then channeled through a nozzle that accelerates them even further. Solid-fueled rockets (like the Shuttle's two booster rockets) contain a putty-like material that contains both oxidizer and fuel. Molded to the entire length of the rocket body, a hollow section at the centre of the mold provides the surface on which burning takes place. Once ignited, a solid-fueled rocket can't be shut down, and neither can its thrust be controlled.

Roger X. Lenard
Principal Member of the Technical Staff
Program Manager, Space Nuclear Systems, Sandia National Laboratories

Nuclear Propulsion for Human Interplanetary Exploration

Nuclear Electric Propulsion (NEP) is enjoying a renewal of interest within NASA. The reason is that, in the near future, NEP is attractive for space science missions – typically one-way trips to the outer planets, or very fast one-way probes for investigating the interstellar medium, well beyond the reaches of the solar system.

Power levels of around 100 kilowatts electrical (kWe) are presently envisioned, providing a payload-carrying capability of as much as 3,500 kilograms (7,700 lb). Space science missions planned for the mid-term future involve sample return systems that spiral out of orbit from Earth's gravity field, travel to Mars, Jupiter, Europa, Saturn, Titan, or more distant objects, depart these lonely outposts and spiral back to Earth with samples. These will need to generate approximately 500 kWe and be able to handle a payload of as much as 10,000 kg (22,000 lb).

A longer-term focus is on human exploration with very high power systems (about 30 MWe), built for speed rather than heavy lift capacity, that would enable round trips to Mars to be accomplished in approximately one year with a short (20–30) day residence on the surface.

Mars is a relatively nearby sentinel in terms of interplanetary distances, only about 50 per cent more distant from the Sun than Earth; still, it is about 50 million kilometres (31 million miles) away. Jupiter, however, is five times, Saturn nine times, and Pluto about 40 times farther from the Sun than Earth. In order to reach these planets within a reasonable time, much higher performance systems will be required. This is because both high accelerations (implying low payloads), high final velocities (implying high specific impulse), and long system life will be necessary. Very high power levels must be assumed because crewed missions will represent a large accelerated mass.

One-year duration round-trips to Mars will require only about one-tenth of the total velocity needed for a trip of similar duration to Jupiter, and the total propellant for such a round-trip can be carried from Earth orbit with minimal penalties. If the spacecraft can be refuelled at the destination, the required initial propellant load is halved. Since the spacecraft has to accelerate only half the initial mass, acceleration time is significantly decreased. However, such trips to Jupiter requires levels of performance implying a specific impulse of at least

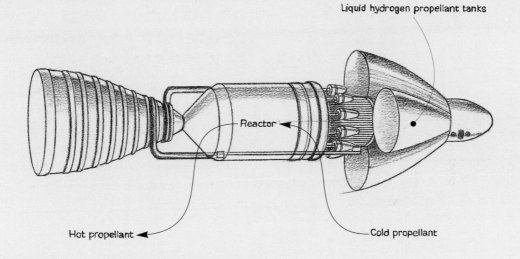

Liquid hydrogen propellant tanks

Reactor

Hot propellant

Cold propellant

20,000 seconds – more than 20 times that of the prototype NERVA nuclear rocket. Reaching more distant objects will require even higher performance.

High accelerations are essential because the acceleration time must be short compared to total trip time, or the use of propellant is inefficient. At a specific impulse of 20,000 seconds, the power system can only weigh 66 tonnes and must produce 3.5 gigawatts of electrical power – if the crew and habitat weigh 30 tonnes, which is about the mass of a proposed Mars crew vehicle.

At present, we can only speculate on the power systems that might produce such power at so minimal a mass. Certainly nuclear power can achieve the necessary specific energies, but scaling up from the best current designs, which typically envision payloads of about 3,500 kg (7,700 lb), will not be an easy matter.

Clearly, much research is necessary to achieve short trip times to planets beyond Mars. One approach would be to increase the time available, but long-term space voyages (especially at zero-g) are still biophysically challenging. One might consider increasing the trip time to perhaps as much as five years, but that simply means that this level of performance is essential to reach Uranus as opposed to Jupiter.

The author and others have examined the requirements to go beyond our solar system using extensions of known technology. Achieving one-tenth the speed of light is essential for interstellar flight, yet this is beyond the energy

NERVA (Nuclear Engine for Rocket Vehicle Application) was NASA's attempt in the 1960s to build a nuclear rocket. Engineers were able to construct and test a prototype twice as powerful as the best chemical rockets before the programme was cancelled in 1971. The NERVA engine was a 'solid core' nuclear rocket and worked by pumping liquid hydrogen into hundreds of tiny channels running through a nuclear reactor. As the liquid hydrogen flowed through these channels, heat from the reactor changed the fuel into a rapidly expanding gas. This gas was then accelerated through an exhaust nozzle to speeds of 35,400 kph (22,125 mph).

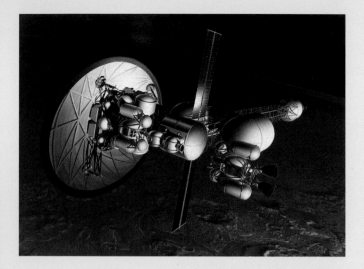

A conceptual nuclear-thermal rocket with an aerobrake disk is shown in orbit around Mars.

density of fission systems and near the theoretical limit of fusion systems. Thus it is necessary to invoke new system architectures – for example, using nuclear fission products as propellant, continuously refuelled reactors, approximately ninety-per-cent-efficient power systems and electric thrusters, an external supply of fission fuels, and magnetic sails to decelerate at the destination. Even so, trip times of sixty years will be necessary to travel 4.5 light years, the distance to Alpha Centauri.

Another alternative is antimatter – particles similar to normal matter but with opposite electric charges. When matter and antimatter come into contact, they annihilate instantly, with all their mass converted directly into energy. However, antimatter is preposterously cost-prohibitive – something like a thousand times the Gross Domestic Product of the U.S. per kilogram!

Given recent advances in genetic research, it is not impossible to speculate that human lifespans will be greatly increased. Thus, even at one-tenth the speed of light, 'reasonable' single-generation trips of 40–50 light years might be feasible. (A session at the 2002 meeting of the American Association for the Advancement of Science was dedicated to debating the concept of multi-generational spaceflight, where family members would be forever encapsulated, living and dying en route to some distant Earth-like planet.)

The only other alternative to such immense undertakings is a breakthrough in physical principles. This might take the form of extracting zero-point energy, some form of 'warp drive,' but we don't yet know how this might be accomplished.

The important point is that NASA is now embarking on a research program to send first one-way probes, then round-trip probes, and finally, human trips to the near planets. The research program will emphasize increased efficiencies and reduced specific mass values. While it may be very hard to lift 4 kg (8.8 lb) per kilowatt of electrical energy (4 kg/kWe), known extensions of available technologies might get to about 1 kg/kWe. These power densities and probable power levels would make human trips to Mars routine and five-year trips to Jupiter realistic. Together with NASA, we are embarking on a very exciting first step.

The solid rocket motors produced 1.55 million kg (3.4 million lb) of thrust to lift the launch system from the pad. The two Titan IV stages ignited and burned out, then the Centaur came into its own. It separated from the second core stage and burned for about two minutes, putting Cassini into a ballistic transfer trajectory for about 17 minutes. Then it fired for the last time and sent the spacecraft out of Earth orbit into solar orbit, toward Venus and the first gravity assist.

The job of breaking away from our planet's gravitational field and putting a spacecraft into space is usually one for chemical rockets. There follows an unpowered ballistic trajectory, sometimes adjusted by smaller rockets during the flight, and/or by solar electric thrusters that use solar energy to ionize xenon gas and force it through a nozzle.

To send large craft very long distances, however, nuclear power will be needed. The U.S. government funded a number of nuclear propulsion projects that included NERVA (Nuclear Engine for Rocket Vehicle Application), a programme cancelled in 1972. Work has continued since then on more advanced designs, albeit at a lower level of effort.

Nuclear rockets are simple in concept – hydrogen gas is superheated as it is passed through a nuclear reactor, producing thrust. The rocket would be assembled in Earth orbit, mated with its payload, eased into a new interplanetary orbit by chemical rockets, then fired up for a flight that could accelerate slowly for years if needed. Fuel rods can be moved in and out of the core, altering the fission reaction and therefore the temperature and thrust.

A useful figure for comparing rocket engines is specific impulse (I_{sp}), a measure of rocket performance equal to the thrust in kilos divided by the flow rate of propellant in kilos per second. A conventional liquid hydrogen-liquid oxygen engine has an I_{sp} of about 450 seconds, NERVA had an I_{sp} roughly twice as great. Future designs such as one fitted with an oxygen afterburner would be still more powerful; a hybrid version can convert the thermal energy to electricity for separate ion propulsion thusters, adding extra capability in mid-course.

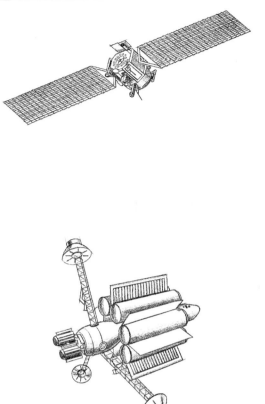

Top: Deep Space 1 was successfully propelled to its encounter with Comet Borrelly by an ion engine. The ion engine does away with the short, powerful blast of a conventional rocket engine. Instead it produces continuous gentle thrust over periods of months or even years by using solar power to accelerate individual ions to very high exhaust velocities.

Bottom: VASIMR (Variable Specific Impulse Magnetoplasma Rocket) is a promising alternative to chemical and nuclear powered rockets for powering a manned mission to Mars.

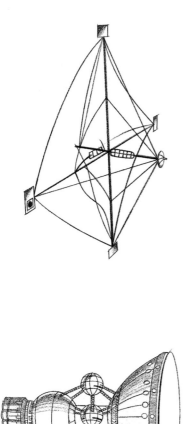

Alternatives include the Variable Specific Impulse Magnetoplasma Rocket (VASIMR). VASIMR is like no other rocket: a model that is only 3 metres (10 ft) long produces an ion stream moving at 300 kilometres per second (670,000 mph). It's a physicist's dream, a throttleable machine that accelerates superheated gases through a magnetic nozzle. In automotive terms, it can lift off in low gear, go into overdrive for the cruise, and shift as needed as it rendezvouses with planets and ultimately returns home.

Amazing feats can be performed by very small solar ion engines once they are in space, free of Earth's gravity. An engine of this kind was built for Deep Space 1, the first mission in NASA's programme for testing advanced spacecraft systems. A gentle flow of xenon gas was solar-ionized, providing a tiny amount of thrust – about the pressure exerted on one's hand by a standard sheet of paper but enough, over time, to increase the craft's speed by an amazing 12,700 kilometres an hour (7,900 mph)! SMART-1, the first European concept for lunar investigations, was designed using a very similar system as its primary mode of propulsion. Electric thrusters based on this principle were first used in spacecraft attitude control systems, a concept pioneered by Russia in the 1970s.

Finally, solar sails, conceived decades ago by Arthur C. Clarke, perform just as their name suggests. At the far end of the spectrum from VASIMR and nuclear rockets, they are powered by the silent but continuous stream of particles in the solar wind, or by the gentle pressure of laser or microwave beams transmitted from Earth.

But to return to the present, what are the most powerful launch vehicles currently at our disposal? The Titan IVB will carry 22 tonnes to low Earth orbit (LEO), or 5.8 tonnes to geostationary transfer orbit; the Space Shuttle 110 tons (orbiter plus payload), but only to LEO.

Europe's star in the heavy-lift stakes is Ariane 5, which from a distance resembles the Titan IV. This system entered service capable of lifting a 6.8 tonne payload to a geostationary transfer orbit, musclepower that was expected to grow to 12 tonnes by 2005, allowing it to launch two satellites at a time.

Top: Solar sails do away with onboard fuel and the mass that goes with it. But they would need to be huge to produce just a little acceleration.

Bottom: Another future rocket without onboard fuel is the fusion powered ramjet engine. Using a vast magnetic field projected in front of it, the ramjet-equipped spacecraft would harvest the hydrogen atoms of interstellar space and somehow force them to fuse.

The Space Shuttle, the size of a DC-9 airliner, combined the glamour of human spaceflight with the efficiency of being able to orbit up to 2.5 tonnes of satellites and spacecraft from one launch, and the boon of being able to reuse every part of the system except the main liquid fuel engine.

The first Shuttle, Columbia, was launched from Kennedy Space Center on April 12, 1981, and glided to an unpowered landing two days later at Edwards Air Force Base, California. Its launch system, like Titan III's, consisted of a liquid-fuel engine flanked by solid rocket boosters (SRBs). But there the resemblance ended. The external tank was truly gigantic, roughly the same diameter as the huge Saturn V first stage. Its engines, burning simultaneously with the SRBs, generated 2.9 million kg (6.4 million lb) of thrust at lift-off. By century's end, various types of Shuttle, some designed to carry astronauts and some to be run completely from the ground, were in development. These included a Chinese variant, Japan's totally reusable robot ship, HOPE, and 'Evolved Expendable Launch Vehicles' from America's rocketry giants.

Such launchers are quite adequate for robotic space exploration, especially as payloads shrink with the advance of miniaturization. When we consider the needs of human spaceflight, however, the picture changes. A system under study by Lewis Research Center, the Magnum, is designed to lift 80–85 tonnes to a LEO staging point for a Mars colonization programme. (It's claimed that an upgraded Russian Energia-B would be powerful enough to lift a whopping 200 tonnes into low-Earth orbit.)

Some of the most innovative and most daring minds in the space propulsion business are exploring the seemingly impossible goals of interstellar travel. Here, time warps and space/time 'wormholes' are being considered by serious scientists who hope to challenge our traditional views of a well-ordered universe and sneak through the back doors of physics to visit the planets of other star systems. Even anti-gravity is being explored, though claims of a successful anti-gravity experiment in Finland were furiously denounced for the simple reason that the notion violates the laws of nature.

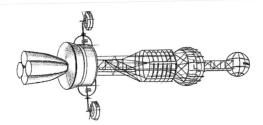

The ultimate rocket propellant is antimatter. Upon annihilation with matter, antimatter offers the highest energy density of any material currently found on Earth. It would take only 100 milligrams of antimatter to equal the propulsive energy of the Space Shuttle. Antimatter, however, is currently the most expensive substance on Earth, at about 62.5 trillion dollars a gram, and it is not easy to handle either!

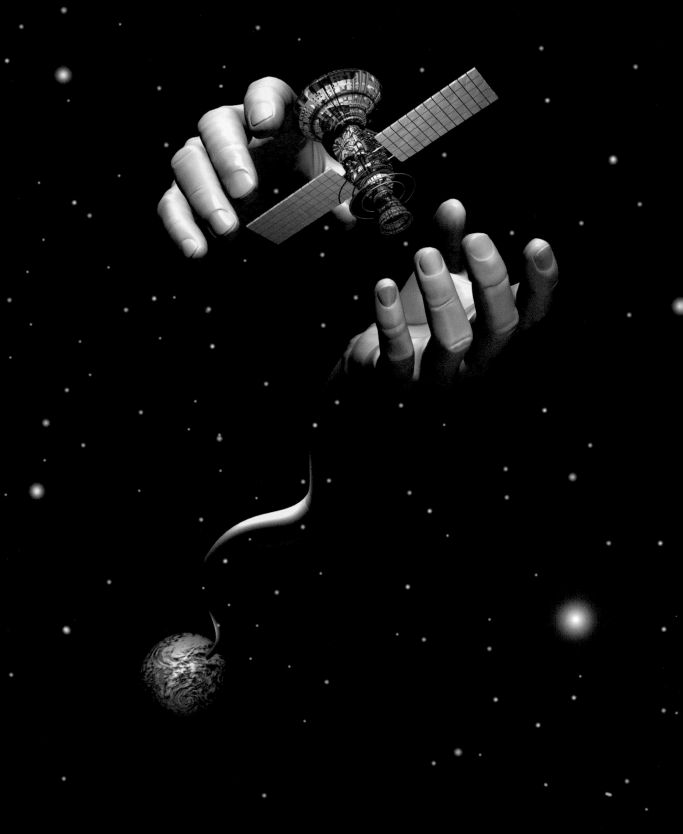

The Billion-Mile Remote Control

'The ship, a fragment detached from the earth, went on lonely and swift like a small planet.'

Joseph Conrad

Modern spacecraft, like those in science fiction stories, have human navigators, plotting their course between and around the objects in the solar system, calculating velocities and arrival times, deciding if and when course corrections must be made. The technology upon which they and their scientist colleagues ultimately depend – radio – is not so very old. There are people alive who remember what life was like before public radio broadcasts began. But think of how far radio took us in the twentieth century! It extended the range of science, and with it the human spirit, to the very edge of the cosmos.

Launch Day, L-5 hours and counting. The gleaming Space Shuttle is poised quietly on the pad, the interplanetary probe tucked safely into its cargo bay, the huge booster rockets pointing majestically into the sky above Kennedy Space Center. Calmly and methodically, engineers are making a wide variety of last-minute checks, from verifying that all electrical connections are complete to making sure that NASA's global tracking system is ready to monitor the flight path. White plumes of vapour stream from the main engine's liquid-propellant exhaust ducts.

At the same time, quiet tension pervades the atmosphere at Johnson Space Center in Houston, where controllers wait to take the helm following launch, and at JPL in Pasadena, where engineers and scientists prepare to track the spacecraft when it leaves the Shuttle a few hours later.

These are the moments of truth. Already innumerable tests have been conducted to ensure the success of this momentous flight. These tests have involved the flight project administration and its contractors, the tracking stations of the Deep Space Network, and the Air Force Eastern Test Range. Every tiny part of the spacecraft, orbiter, launch vehicle, and ground systems must perform exactly as it is programmed to perform – there can be no turning back after ignition save the embarrassing alternative of bringing the Shuttle back from its orbit early.

In a painstaking process, larger and larger portions of the spacecraft have been combined and tested, then completed and subjected to heating, cooling, and vibration to make sure they can survive the environments of launch and space. Mission personnel have trained intensively to meet the task ahead, so that their capabilities are honed to a sharpness that will guarantee the success of a mission that will never have a second chance.

L-2 minutes and counting. The last moments of countdown are almost wholly devoted to the rocket systems – pressurising the liquid hydrogen tank, starting the solid rocket booster, hydraulic power units and gimbal tests, starting the main engine at L-6 seconds and, at L-0, the SRBs.

Lift-off! The huge vehicle rises slowly into the sky, gradually gaining speed as the giant liquid and solid rockets lift its great mass against the Earth's gravity. Even outside the Launch Control Center three miles away, spectators are buffeted by sound waves, their clothing ruffled by acoustic energy.

After the Shuttle's orbit is established, the crew can deploy the spacecraft, easing it out of the cargo bay and into space. After the Shuttle has moved to a safe distance, the spacecraft's detachable engines ignite, taking it from Earth orbit to a solar orbit that will carry it on a new mission of discovery.

In this illustration, based on the Galileo launch, the spacecraft is able to fit in the Shuttle cargo bay, and the launch sequence is somewhat different from that used with unmanned systems like Ariane or the Titan IVB/Centaur that put Cassini into orbit. The differences in the run-up to the flight and its first few hours are striking; afterwards, the similarities are striking.

Let's back-track and take a look at what orbits actually are, and how they are achieved.

Most people know that, if you give an object a great enough boost away from Earth, it will overcome the pull of gravity and not fall back to the surface. A typical spacecraft follows a ballistic trajectory after its rockets cut out, just as if it were a ball thrown up from Earth's surface. Provided the apogee (highest point) of this trajectory is beyond atmospheric drag and the spacecraft is travelling fast enough, it will slip into a path where the element of its own inertia that tends to carry it further away from Earth is precisely balanced by Earth's own gravity. This is a circular or elliptical orbit,

However, if a spacecraft is launched at a high enough velocity, it will not be able to maintain even an elliptical Earth orbit. The apogee becomes infinite, the curve of its orbit turns into a parabola, and the vehicle escapes Earth's gravitational field. At this point the Sun's gravity becomes the dominating force and the vehicle settles into its path around the Sun. The velocity required for this to occur – escape velocity – is 11.18 kilometres per second (roughly 36,000 feet per second, or 25,000 miles an hour).

With each increase in velocity, the orbit becomes more and more elongated, so that the spacecraft can rendezvous with a planet, take measurements close to the Sun, peer at an asteroid, or take a chunk out of a comet for analysis.

Spacecraft bound for interplanetary flight get a 'leg up' through use of an idea credited to Walter Hohmann (1880–1945) of Essen, Germany. Hohmann's outstanding work was a mathematical study of interplanetary trajectories, and to this day his name is immortalized by the term 'Hohmann transfer,' which describes a means of changing from an Earth parking orbit to a round-the-Sun orbit. Through its use, or the use of a more modern ballistic transfer trajectory nicknamed the 'Porkchop,' the spacecraft is given just enough velocity to reach its target planet while following an elliptical orbit around the Sun.

With this technique, the final solar orbit is reached by what almost amounts to a second launch, this time from orbit. Take the case of the Cassini spacecraft. It was first established in Earth orbit, then the Centaur upper stage ignited at just the right time to break the spacecraft away from Earth's gravitation, so that it would 'fall' to its new orbit around the Sun. Then the Centaur detached itself and Cassini was on its own. The new orbit would take the spacecraft to Venus in the first leg of its long journey to Saturn.

Cassini passed by Venus in April 1998, getting a gravity assist that added to its energy and increased the size of its orbit. That was just the first of four gravity assists, each of which results in the spacecraft trading energy (angular momentum) with the planet it encounters. The spacecraft received a huge boost each time, simultaneously slowing the planet by an amount so tiny that it can't be measured. After orbiting the Sun again, it revisited Venus for another gravity assist, and then soared back to Earth. At this point, Cassini gained enough energy to reach Jupiter, and a final gravity assist for the home stretch. By design, upon its arrival at Saturn, Cassini would use up a large portion of its remaining propellant by firing its main rocket engine and braking into orbit around the ringed planet. That's the most ambitious interplanetary pinball scenario attempted by space science.

The task of planning such a planetary encounter is enormously complicated. The planetary orbits represent extremely small, fine paths compared with the hugeness of the solar system. And a planet may arrive at a

The Hohmann transfer was adopted as an energy-efficient method of transferring a spacecraft between two non-intersecting circular orbits. When transferring from a lower to a higher orbit (top), the spacecraft performs a velocity increasing burn (1), which pushes it into a transfer orbit. The increase in altitude reduces the velocity of the spacecraft, so when it reaches the altitude of the upper orbit it performs another velocity increasing burn (2) to circularize its orbit. Without the second burn, the spacecraft would remain on the elliptical transfer orbit (3).
When transferring from a higher orbit to a lower orbit (bottom), two burns are still required (1 and 2), except this time they are velocity *decreasing* burns. The first burn puts the spacecraft onto its transfer orbit. But as its altitude decreases it gains velocity, so a second burn is required to circularize the lower orbit and prevent the craft from remaining on its transfer orbit (3).

given position on that orbit only once in several lifetimes – every 248 years in the case of Pluto.

In all space missions, trajectories must be analysed by taking into account a multitude of variables – for example, launch timing and direction, mission duration, and the orbits and orbital planes of the outer planets.

Despite the precision with which these trajectories are planned, unknown conditions – ranging from meteoroid impacts to inaccuracies in planetary ephemerides – make midcourse corrections and orbital trims necessary from time to time. These problems are revealed when data from NASA's tracking system show undesirable changes in velocity (this is the now-familiar delta-vee [ΔV or DV], the Greek letter delta being the accepted scientific shorthand for change or difference, and the V standing for 'velocity'). Thrusters, or on-board rocket motors, must then be activated to bring the spacecraft's speed and direction back to the desired values.

Similar effects of the space environment will cause the attitude or tilt of the spacecraft to change, disrupting both the alignment of its antennae with Earth and the orientation of on-board scientific sensing equipment. Communication can be disrupted if the spacecraft shifts off its alignment with Earth by as little as five-hundredths of a degree.

In other words, the spacecraft must be able to position itself so that parts of it can point in a particular direction at the same time that it is moving along its trajectory. Consider the way we humans are engineered. When we look to left or right, our eyes scan the x direction; when we look up and down, we scan the y direction; when we nod our heads from side to side, our entire head moves in the z direction. If we want to watch a tennis match, we use a complex set of large muscles to get us to our seat at the tennis court, and a set of smaller muscles to move our heads so we can see and hear the action to best advantage. Our adventuring robots are like robotic heads in space, endowed with most of the senses we have in our own heads plus a few more. And they too must be able to move. As is the case with other spacecraft, Cassini does this with its attitude and articulation control subsystem (AACS).

The Cassini AACS is larger and more powerful than Galileo's for the simple reasons that it is a bigger spacecraft and a longer mission. Because of its long cruise time (6.7 years) and lengthy time in orbit around the Saturnian system (4 years), it had to carry a large amount of propellant for in-flight trajectory-correction manoeuvres and attitude control.

Cassini's 'leg muscles' are two rocket engines (one a backup) that burn nitrogen tetroxide (N_2O_4) and monomethyl-hydrazine (MMH) to make adjustments to the flight path when needed. These are hypergolic fuels; that is, they ignite spontaneously when they come together. The main rockets have a thrust rated at 445 newtons (10 newtons is roughly the force exerted by a weight of a kilogram (2.2 lb) at Earth's surface). The engine fired for 90 minutes in December 1998, to put it on the correct course for a 1999 flyby of Venus; its next 90-minute firing will ease it into orbit around Saturn in July 2004.

Sixteen much smaller systems, thrusters that move the craft by puffing out tiny amounts of hydrazine (N_2H_4), are used to change its attitude (i.e. 'move its head') so that it is pointing in the right direction at the right time. These are each rated at a minuscule 'baby's breath' force of 1 newton.

Finally, the spacecraft has reaction wheels to make very precise attitude adjustments when it has arrived in Saturn orbit, and is making measurements of Saturn, its rings, and its moons. When the spacecraft has to be rotated in one particular direction, one of these three wheels is spun in the opposite direction by an electric motor. Like rockets and thrusters, they operate according to Newton's third law of motion, which states that 'to every action there is always an opposed and equal reaction.' Two or even all three reaction wheels may be spun up to accomplish complicated manoeuvres.

The AACS's first job was to get a 'fix' on the Sun, right after the spacecraft was freed from the launch vehicle. Throughout the mission it has to point the appropriate radio antenna toward Earth, or toward the Huygens probe during its brief descent to Titan. On command it must turn the craft for trajectory-correction manoeuvres, science observations, and science instrument calibrations. Among its many other tasks, it has to steady the spacecraft for special experiments such as gravity wave measurements. (On its way to Saturn, scientists want to monitor Cassini's distance very precisely in the hope of detecting passing gravity waves – theoretical ripples from violent cosmic events which could distort the space between Earth and the spacecraft and cause it to contract or expand.)

To perform these tasks, the AACS is equipped with a super-smart set of computer-compatible sensors. A stellar reference unit determines exactly which way the instruments are pointing and indicates what changes are occurring in the reaction wheel system. Inertial reference units serve as a backup by registering every tilt and turn of the spacecraft during its long journey. Special electronics control main engine and thruster power for periods ranging from 1 second to 3 hours. An accelerometer, designed to withstand 15,000 hours of on-off operation over five years, determines velocity changes and turns off the main engine at the right time. Engine gimbal actuators keep the main engines pointed in the right direction during up to 200 'propulsive manoeuvres.'

The craft is a space robot that is invested with the equivalents of eyes, ears, voice, and muscle. Each spacecraft takes with it the hopes and dreams of thousands of scientists and engineers and, most importantly, the special sense of wonder and imagination that is so great a part of human nature.

Whispers from Space

Signals from any outer-planets spacecraft must travel many hundreds of millions of miles – from a speck in the solar system to the globe-circling antennae operated by NASA and ESA. By the time it arrives, the signal will have become incredibly weak.

These minute radio signals do a near-immaculate job of sending commands to spacecraft millions of miles away, and receiving information back via the marvel of telemetry (from the Greek for measurement at a distance). At the speed of light, it takes about eighty minutes for these signals to travel to or from Saturn; by comparison, signals get to the Moon in just over a second.

Heinrich Hertz (1857–94), who discovered radio waves in 1887, and Gugliemo Marconi (1874–1937), who sent the first successful transatlantic radio message in 1901, would have been delighted.

Here's a very quick review of how Cassini's information gathering system works. On board the craft, scientific instruments collect images, spectra, and other data, along with engineering information on the spacecraft's state of health (including its position, orientation, and velocity). The spacecraft stores and transmits these data to Earth receiving stations. This information is moved to JPL (or the European Space Operations Centre [ESOC] in Darmstadt) via land lines, communications satellites, and microwave links. Engineering data go to mission controllers; science data are processed and sent to the scientists who are running that particular part of the mission. Eventually the scientists present and publish findings to the scientific community; some information finds its way to science journalists who pass it on to the general public.

The greater the distance, the weaker the signal. If you are transmitting 20 watts of 'S-band' power from space, using a 4.2-metre (14-ft) antenna dish, you will receive only two-millionths of a billionth of a watt from 80 million kilometres (50 million miles) – and this is using one of the DSN's three very powerful 63-metre (210-ft) receiving antennae! At 640 million kilometres (400 million miles), the same signal will decrease to about one-sixtieth of this value.

Incredible as these figures may seem, they are well within the reach of NASA's 70-metre (230-ft) antenna at Goldstone, California. Its lower limit of signal detection is one one-hundred-thousandth of a millionth of a billionth of a watt (1×10^{-17} milliwatts)! JPL's Fred Siegmeth explained:

'It's a fantastically small amount of power. If one would drive a 100 per cent efficient electric motor with a power of 10^{-17} milliwatts it would take 30,000 years to lift a weight of one milligram to a height of one millimetre (0.04 in)! By way of comparison, an aspirin pill weighs about 100 milligrams.'

When Voyager reached faraway Neptune in 1989, the radio signals were indeed tiny – a quadrillionth of a watt, twenty billion times less than the power needed to run a digital watch! To catch this incredibly small whisper from space, NASA's three largest radio antennae were enlarged from 63 metres (210 ft) in diameter to 70 metres (230 ft), and their 'hearing' was made sharper by installing new amplifiers that would run at very low temperatures, reducing the amount of background noise with which the receiver must contend. More antennae were hooked into the world-circling network, including Japan's 63-m (210-ft) Usuda Radio Observatory and the Very Large Array, a cluster of twenty-seven 25-m (81-ft) antennae operated by the National Radio Astronomy Observatory in the remote San Agustin Plain west of Socorro, New Mexico. In all, thirty-eight antennae all over the world were networked together for the Neptune encounter.

Information is transmitted to and from the spacecraft by microwave – essentially the same basic type of system used on Earth to beam telephone and television signals across wide spaces where cables and fibre-optic links are uneconomical. As the name suggests, microwaves are very short radio waves with higher frequency than the 'ordinary' waves used in commercial radio transmission. Whereas broadcast radio frequencies are measured in thousands of cycles per second (kHz, or kiloHertz), microwaves are in the billions of cycles per second (GHz, or Gigahertz) region. Their great advantage is that they can carry more information than longer wavelengths; their disadvantage (in Earth applications) is that they travel along a line of sight and thus must be relayed over horizons and around mountains and other obstructions.

The Voyager spacecraft utilized two microwave frequencies – around 2.3 GHz (S-band), transmitted at 6.6 watts, and 8.4 GHz (X-band), transmitted at 21.3 watts. Since communication occurs on a two-way basis, the direction is generally called either the uplink (Earth to spacecraft) or downlink (spacecraft to Earth). Cassini's X-band downlink is transmitted at 7-8 GHz, generally at less than 10 watts.

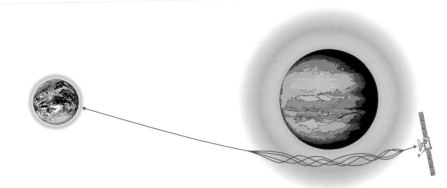

Measuring the refraction or distortion of various radio frequencies can be used to estimate the density, pressure and composition of planetary atmospheres.

It is possible to send data to Earth on the X-band frequency six times faster than on the S-band link. But the S-band may still be the better system in some cases because it is neither so susceptible to weather conditions at the receiving site nor so sensitive to interplanetary conditions such as the radio noise generated within the target planet. Under some circumstances, two or three of the downlink transmitters may be operated at the same time, either to improve the quality of the signals received or to measure planetary fields by comparing the distortions affecting the different frequencies.

The Cassini orbiter communicates with Earth via a high-gain antenna (HGA) developed by the Italian Space Agency. Two low-gain antennae (LGAs) are also used to send data and receive commands. The LGAs amplify the 'voice' of the spacecraft to that level which is best under the circumstances without draining more power than necessary, and without requiring the pointing precision and directivity needed for HGA operations.

Data are translated into binary code (see p.135) and handled on board the spacecraft by three basic methods:

- Real time, in which they are transmitted directly to Earth without storage:
- Telemetry store, in which they are stored on-board in a 'buffer' at the system's own speed, but transmitted to Earth at a more suitable bit rate; and
- Memory readout, in which they are stored on-board and transmitted later on command from Earth (or from the stored sequence) at any desired bit rate.

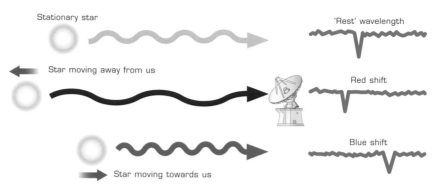

The Doppler effect enables us to work out the speed and direction of objects relative to ourselves. For example, a star moving away from us would have its light waves stretched out, becoming redder ('red shift'), while the light waves from a star moving towards us would be compressed and the light would appear bluer ('blue shift'). This Doppler shift (as it is known) applies to all electromagnetic waves across the spectrum.

That's the story of how information gets to and from the spacecraft. But we're not through with radio, which can give us information about planetary environments even while transmitting data to Earth. 'Radio science investigations' can measure atmospheric pressure, density, temperature, and composition at different altitudes. At Saturn, for example, a special Ku-Band (10-18 GHz) radio will scan Titan's hidden surface from the Cassini spacecraft, and S-band, X-band, and Ka-band (20-30 GHz) signals will probe the secrets of the planet's rings.

Some of these experiments are based on refraction – the radio beam is bent by the atmosphere, just as light is bent when we look at an object through water or a piece of quartz. Observations of how much the radio beam is refracted can, for example, provide a measurement of atmospheric density. The denser the gas, the more the beam strays from its straight-line path.

Doppler-effect measurements are also important radio science tools, revealing how much 'pull' is being exerted on the spacecraft by gravity. This changes the spacecraft's acceleration and therefore creates a Doppler shift in the frequency of the radio signal received on Earth. Years ago, Doppler experiments carried out during the Mariner flights showed the mass of Mars to be 0.11 Earth masses. By a similar analysis, Venus was determined to have a mass equivalent to 0.82 Earths. Data from the lunar orbiters likewise revealed the presence of extra-dense material (mass concentrations or mascons) under the Moon's surface.

At Jupiter, measurement of Voyager's radio signal revealed the distance of the spacecraft to within a few metres, and its speed to within one half millimetre (0.02 in) per second. Voyager's radio Doppler effect was also used to 'weigh' Neptune and Triton.

Michael Faraday suggested that there were such things as electromagnetic fields in 1832; James Clerk Maxwell predicted that radio waves existed in 1873;

Gugliemo Marconi sent the first radio message across the Atlantic in 1901; the first commercial radio station started operation in 1920. Today we are able to gather radio whispers from the far edge of the solar system, and from light years beyond. In the words of Arthur C. Clarke, 'Any sufficiently advanced technology is indistinguishable from magic.'

Managing the Data Links

As Mariner 2 sped toward its encounter with Venus in 1962, NASA scientists were shocked to learn that there had been an error during the spacecraft's departure from its initial orbit around the Earth. Using a computer to chart its flight path from radio data, they learned that Mariner would miss its target by more than 400,000 kilometres (250,000 miles).

The spacecraft was coasting in free flight, on a path that would eventually decay into a useless orbit around the Sun. Even then, NASA engineers were ready for such an emergency. Because they knew only a miracle could give them an initially perfect aim toward Venus, they had provided the spacecraft with a small rocket engine to correct its trajectory as needed. Again using computers, they determined what amount of velocity change (ΔV) would bring the trajectory up to par – 51 metres (170 ft) per second. They then radioed a set of commands to correct the course by applying thrust in a set direction for a set amount of time. The thrusters came to life, made the necessary change, and Mariner 2 continued on, becoming the first spacecraft to encounter Venus.

Navigation and control technology improved steadily during the 1960s until, by 1969, a ΔV of only 4 metres (13 ft) per second was required to correct the initial target miss of 26,900 kilometres (16,700 miles) of Mariner 7 on its way to Mars. As computers became more powerful and navigators more experienced, these trajectory trims became steadily smaller and fuel requirements became lighter, making more space for science experiments.

Credit for these improvements was due to the increased sophistication and accuracy of the launch vehicle guidance system, on-board satellite control systems, and the improved accuracy of NASA's tracking and orbit determination system, the Deep Space Network (DSN).

The DSN performs four basic functions – tracking, data acquisition, command, and control.

- **Tracking** – locating the spacecraft, calculating its distance, velocity, and position, and following its course;

- **Data Acquisition** – recovering telemetry information from the spacecraft on its general condition and the scientific data collected;
- **Command** – sending signals to guide the spacecraft flight and operate on-board scientific and engineering equipment; and
- **Control** – on-the-ground decision making with regard to flight and ground station operations.

The DSN is served by a worldwide chain of stations. These include major antenna sites in the United States (four telescope at Goldstone, California), Spain, and Australia. They are placed about 120 degrees apart around Earth's circumference, so that contact can be passed from one to another as our planet revolves on its axis. Most antennae are 34 metres (113 ft) in diameter and some receive at Ka-band as well as X-band. Others will receive S- and X-Band, three of which measure 70 metres (230 ft) across. A similar tracking network, ESTRACK, serves ESA spacecraft.

The original Goldstone 63-metre (210-ft) antenna was dedicated on April 29, 1966, following a six-year construction project. Tall as a twenty-one-storey building, it weighs 2,500 tonnes. At the time of completion it was the world's second-largest fully steerable antenna, after the University of Manchester's Jodrell Bank radio telescope in Cheshire, England. But size is not everything – the Goldstone antenna, with its more powerful receiving and transmitting capacities, is considered the world's most capable tracking station.

Steering the Interplanetary Flight

NASA's Deep Space Network operates through the Space Flight Operations Facility (SFOF) at JPL. This is the control point for all unmanned U.S. space flights, from which project managers, flight operations personnel, spacecraft engineers, and teams of scientific experimenters are kept informed of the mission's status.

The SFOF communicates with the Deep Space Network stations via a direct computer link in addition to telephone and e-mail, routing spacecraft data to a specialized processing system that converts the coded spacecraft signals into readable form. This information is then fed to other personnel within the SFOF to keep them up to date.

Monitoring and control of ESA spacecraft is handled by the Mission Management and Control Centre (MMCC) at ESOC in Darmstadt. The MMCC is

Donald L. Gray
Voyager Navigation Team Chief
Jet Propulsion Laboratory

Navigating to Neptune

Spacecraft navigation is a test of innovation, accuracy, and the wise use of propellant, managed by teams that cooperate with scientists, spacecraft engineers, and communication engineers.

From my point of view, the greatest excitement of unmanned spaceflight springs from the innovative and surprising ways that mission-critical problems are solved. These challenges often go unnoticed by a public that is justifiably enthralled by the results of each successful mission. As examples, let's look at a few war stories from the Voyager Project, which in the 1970s and 1980s provided our first close-up looks at Jupiter, Saturn, Uranus, and Neptune.

Neptune as it would appear from a spacecraft approaching Triton, Neptune's largest moon. This computer-generated montage uses images of Neptune and Triton taken by Voyager 2.

Voyager looked just fine until the first manoeuvre, when we learned that the thruster jets were hitting struts on the spacecraft superstructure. The result was a potentially disastrous 20 per cent reduction in ΔV (velocity change) capability. The only way to save much ΔV would require doing the planned post-Jupiter timing correction closer to the planet flyby than planned, but a full manoeuvre would interfere with science observations. Our final solution was to gain maximum effect from the flyby as the direction of the spacecraft's velocity was sweeping past the Earth. With the spacecraft positioned so that no change in its attitude was needed, the thrusters were turned on and a powered flyby was accomplished with a single command. The science team helped by using a less costly timing change en route to Saturn. These changes reduced the propellant cost of this manoeuvre from 67 metres (223 ft) per second to 14 metres (47 ft) per second, saving enough ΔV to complete the mission.

The next problem occurred when on-board computers determined that the communications uplink had gone awry and switched to an alternate receiver – one that was, alas, totally dead. After an agonising week in which no commands could be sent to the spacecraft, the system automatically returned to the original receiver, which could no longer search for our signal frequency because a capacitor had failed. We were able to solve this problem by varying our transmitted frequency in such a way that the spacecraft would receive a nearly constant frequency.

Then the gears that turn the scan platform, on which many of the science instruments are mounted, became stuck during the Saturn encounter period. The solution was to put the scan platform in the equivalent of a car's low gear.

By slowly and cautiously grinding away in low gear, whatever was blocking the gears was finally removed.

Uranus and Neptune posed new challenges. Neither the gravity nor the positions of these distant planets was well known at the time, and it takes a radio signal four hours and ten minutes to travel from Earth to Neptune! Several tools had been developed to deal with such challenges, and had been used successfully in the encounters with Jupiter and Saturn. Now they became even more crucial to success.

Uranus lies on its side, circled by its five known moons. Seen from the spacecraft, it looks like a large target. But when we received them at JPL, our computer said the images of those moons showed somewhat incorrect rates of motion about Uranus. Analysis of the moons' orbits showed that Uranus's gravity was 0.26 per cent greater than estimated from Earth. This meant that the first approach manoeuvre had to be redesigned. It was 343 kilometres (213 miles) farther out from the original position, so that the greater-than-expected gravity would fling the spacecraft on its correct trajectory to Neptune.

Unfortunately, if we did the last pre-Uranus manoeuvre there would not be enough tracking data afterward to ensure best-possible science observation timing and direction. In this case the magic trick was to depend on the navigational tools that we had on hand, and deliberately not do the last manoeuvre. Doing so made more tracking data available to support late updates to science observations, as well as simplifying encounter operations. The resulting encounter science was spectacular.

Neptune was a very different encounter. Uranus's multiple moons had made it possible to assess the optimum distance for the flyby at first approach. But the only known moon at Neptune was Triton. We needed to find a yet-undiscovered inner moon, and then to refine its orbit estimate enough to use it for navigation. Sure enough, the new moon was found, just in time, and used to obtain an accurate encounter.

Neptune is not the end of the story. Both Voyager spacecraft (along with Pioneer 10 and Pioneer 11) continue to depart the solar system, seeking the tenuous boundary where our Sun's radiation is overcome by galactic radiation. The radio signal to Pioneer 10 for example takes just over 20 hours to get there and back, which means you could send a signal at 2 p.m. one day, go home, have a good night's sleep, and come back at ten the next day to hear the reply.

responsible for both satellite and payload, including acquisition of telemetry and uplinking of telecommands. The MMCC runs software systems and is also responsible for calculating orbits and assuring optimal flight performance.

The Cassini Real-time Operations Element is one of several teams that will be hard at work at any one time in the SFOF. Immediately following launch, mission operations are handed over to these experienced in-flight specialists, who have been thoroughly schooled in the features and function of individual subsystems. Training includes exercises and operational readiness tests designed to make sure that team members know their job duties and how to perform them with excellence. It also provides an opportunity to put spacecraft hardware and software through their paces and make sure that written procedures are right for the job.

Cassini's mission controllers work around the clock, each taking the name 'Ace' as his or her shift begins. Ace, whomever he or she is, is responsible for making sure that data are routed to the right destinations; that the system navigators are supplied with accurate tracking information; that the spacecraft has been given the proper computer commands; and that precious telemetry data are stored and delivered to scientists and spacecraft engineers.

Ace monitors computer workstations that display the spacecraft's general state of health and the real-time status of the network of interlinked antennae and systems that provide communications between the DSN, co-operating ground sites, and the spacecraft.

If it appears that the mission has a problem, Ace immediately takes action to correct it. A swift reaction can be vital, but fortunately problems are rare.

Data on power supplies, fuel tank temperatures, radio signal strength, gas tank pressure, and rotation rate are included in the wealth of engineering information that comes in on the downlink. Ace has to make sure this information is processed and routed to the right people. Then it's up to these specialists to handle spacecraft problems, like software glitches, if and when they occur.

Raw telemetry data may be an incomprehensible jumble of numbers. It may be meaningless to us, but the computer recognizes its meaning and translates groups of numbers into engineering values. They emerge from the computer in readable terms.

These data are carefully scrutinized by engineers and scientists who staff SFOF working teams. These differ for each program, but typically one might be the Flight Path Analysis group, which studies tracking data on the spacecraft's flight path; a second, Spacecraft Performance Analysis, which keeps tabs of measurements from on-board engineering equipment: and a third, Space Science Analysis, responsible for data received from scientific instruments.

The work of these teams is crucial to the success of the Cassini mission – and to other spacecraft as well, since ordinarily a number of space flights are monitored simultaneously.

A couple of events from November 1998 illustrate the kind of work processed at the SFOF. On November 10, Deep Space 1's ion propulsion engine shut off, soon after it had been turned on for the first time, and without apparent reason. On November 21 and 22, Galileo unexpectedly reset part of its computer system twice, causing photo opportunities to be lost at Europa.

With no technician up there to poke around and see what was amiss, SFOF personnel millions of kilometres away on Earth have to check what data they have, provide theories on the cause of the problem, and if possible do something about it. Sometimes it is impossible for them to do anything but cross their fingers and hope for the best. And in November 1998 the best did happen. Deep Space 1's engine was restarted successfully, and it was concluded that some sort of contaminant must have temporarily lodged in the engine's throat. Galileo righted its own problem too, after what the team believed was a case of spacecraft-type radiation sickness – rays from Jupiter's environment interfering with the operation of electronic circuits.

But the SFOF is not all about problems. Most of all, it is about the marvellous successes that emerge from these delicately engineered, complex missions, and the wonders that they bring back to Earth. 'The wonderful thing about planetary encounters was that sense of immediacy, of being present at the very moment that scientific discoveries were being made,' said George Alexander, head of public affairs at JPL, after he viewed the first-ever outer-planets images with a roomful of journalists. 'It was truly thrilling to sit in the JPL news room and watch the monitors suddenly fill with the first images of far-distant shores.'

Navigation and Control

Like any terrestrial voyager, our spacecraft must stay on course so that it reaches a given destination at a given time. The tiny craft must operate virtually without error despite the passing years and the ever-lengthening communications time lag with Earth. It must take note of many different reference points in space. And it must operate without human command as well as technology will permit.

Making sure that our spacecraft conforms to its strict timetable depends on two main factors – navigation and control. Navigation requires measuring the spacecraft's position and velocity with respect to its flight plan. Control involves modifying the trajectory when navigational errors are detected, using the thrusters or rocket engines described in Chapter 4.

Three separate phases of the control routine occur between ignition and planetary encounter – launch guidance (to the time when escape velocity is reached); midcourse guidance; and approach guidance. Each requires a somewhat different mode of operation.

From the moment it leaves the pad, the spacecraft is aimed toward a zone where its trajectory will intercept either the target planet or a gravity-assist planet. Some course corrections can be made early in the flight, so that the spacecraft will at that moment be aimed precisely for the target zone. External influences like the solar wind and radiation pressure, however, will deflect it from the course so that more corrections are necessary later in the flight.

Such corrections require instructions to the spacecraft to first point its propulsion system in the right direction (i.e., roll 8.21 degrees and pitch -130.38 degrees) then increase velocity by a certain number of metres per second with thruster energy. After this operation is completed, the result is confirmed by DSN or ESTRACK communication with the spacecraft.

Midcourse corrections are carried out mostly by external control; in other words, with detailed 'command loads' sent from Earth to the flight computer. These ensure that the spacecraft will travel a path within certain well-defined limits. Straying beyond this limit will require more corrective thrust than the on-board propellant supply can spare.

Where's that Spacecraft Now?

The spacecraft is no more than a mote of dust in the almost incomprehensible vastness of space. No telescope could possible search it out. So just where is the spacecraft, and how fast is it moving? The answers lie in precise tracking, which relies on radio but is quite different from the telemetry used for carrying engineering and science data. It involves measurements of the downlink's Doppler shift and the angles at which DSN antennae are pointing.

Doppler shifting of the downlink occurs because the spacecraft is moving relative to the receiving station. If it moves towards Earth, the frequency of the signal reaching Earth is increased, while if the spacecraft moves away, the frequency is reduced. Data derived from frequency information is computer-processed to remove the effect of Earth's rotation and orbit, revealing a 'true' Doppler shift that indicates velocity to a very high degree of accuracy.

Distance measurement can be accomplished by adding a coded 'ranging pulse' to the uplink to a spacecraft, having the spacecraft return the pulse on its downlink, and recording the elapsed time for the round trip. This information is computer-massaged on the ground at the DSN to eliminate Doppler effects and other artefacts such as the time it takes for the pulse to move through electronic components, and the distance measurement results.

Spacecraft position can be determined by angular measurement, which is done by comparing information from two DSN stations located on separate continents. Since they are a considerable distance apart, they constitute what is called a very long baseline for the measurement – thus the technique is called very long baseline interferometry, or VLBI. With recording instruments running, the two stations first point at the spacecraft signal, then move to a quasar, a distant object whose position is known with great accuracy. They then move back to the spacecraft and stop recording. The data are compared to compute the spacecraft distance and position by triangulation. The results are very precise, requiring that the positions of the stations at each end of the baseline are known to an accuracy of 3 cm (1.25 in).

As the spacecraft approaches its target planet, the approach pathway narrows, like an elongated, invisible cone in space. At its apex is the final aiming point, which is the best-possible computed position for achieving planetary encounter with the expected amount of gravity assist.

Insertion into planetary orbits, and activities carried out during those orbits, demand immense care, for there's never a second chance. The same applies to gravity assists. In the Voyager 2 mission, if the target zone at Uranus had been missed by just one kilometre (0.62 mile), this would have meant a 14,000-kilometre (8,680-mile) miss at Neptune!

Not many manoeuvres have been as critical as Galileo's Jupiter Orbital Insertion, which had to start at 27 minutes past midnight on December 8, 1995, under control by signals that had to be transmitted from Earth precisely 52 minutes earlier. If things went wrong, the spacecraft would have flown past Jupiter and into solar orbit. First, to ensure stability, the orbiter was spun up to 10.5 revolutions per minute (spacecraft, like bullets, fly straighter when they spin). Then the 400-newton engine went into a 49-minute burn, slowing its speed by 643 metres per second (1,438 mph). That – with perfect DSN tracking and precise navigation – was exactly what was needed to put Galileo on its seven-month first orbit around Jupiter.

During the long cruises to and between the outer planets, command loads of engineering information can be uplinked at a fairly leisurely pace, some of them being valid for weeks at a time. It's a different story during close approach to a target planet, when one command load, full of rapid-fire instructions to many parts of the spacecraft, especially to the science experiments, might zap through the system in a few hours.

Things get so exciting, and decisions so critical, at this time, that pre-encounter training is required, much as was the case in the pre-launch preparations. The flight systems and ground team put the spacecraft – and themselves – through a series of exercises that simulate encounter. The system is subjected to realistic challenges, such as changes in data rate and format, and spacecraft manoeuvres. The goal is to sharpen mission readiness, prove out systems and procedures, and reveal flaws or weaknesses that need correction.

With earlier space probes, when shorter distances were involved, all navigation was handled by radio command after the spacecraft's position was computed at Earth installations. But for missions to the outer planets, it soon became obvious that automatic measurements of planet direction would be 'a practical necessity' so that at least part of the navigation and control tasks could be handled by the spacecraft itself.

In all, there were four main reasons for this conclusion:

- Calculations made on Earth may be accurate enough in terms of the information available but might be rendered ineffective by small, unexpected forces acting on the spacecraft;
- Systems switched on and off by radio command from Earth require a certain minimum correction capability, meaning that more weight might have to be allocated for thruster systems;
- Radio system performance is not entirely predictable (halving the signal quality could cause 10 to 20 per cent of 'Grand Tour' missions to miss reaching their final planets); and
- On-board planet direction finding would greatly reduce the need to rely on relatively inaccurate written ephemerides, meaning that trajectory corrections would be smaller and propellant requirements could be reduced.

Telemetry and computing have become such fine arts since the TOPS era, when this list was compiled, that the urgency of these issues has softened. Nevertheless, the flight computer is charged with thousands of 'generic' software routines that can be activated automatically or through instructions contained in a command load. Cassini is set up to recognize more than a thousand different commands, sent on the uplink by a computer that varies the phase of the radio signal, modulating it with pulses.

Back on Earth, after mission planners have explained what they want to accomplish, a sequencing team converts their plans into commands that make the spacecraft carry them out. Special software is used to select the data, place this information in the proper time order, make sure it doesn't interfere with some other requirement, and confirm that the spacecraft has memory to accommodate it. These command data, which relate to a given time period, are called a sequence or command load. Some 'real-time' commands need to be uplinked during previously scheduled operations. These are usually shorter than the loads delivered by the sequencing team. They may be originated by anyone who has an urgent need to get a message through to the spacecraft, such as scientists who need to make quick adjustments to their instruments.

In 1962, one Mariner spacecraft bound for Venus was lost because a hyphen had been left out of the instructions by which a computer controlled the launch vehicle. The Atlas-Agena launch system began to veer off course and had to be destroyed for safety reasons. Fortunately our computer systems have improved immensely since then – though great care is taken to make sure that they are working correctly before launch. Lift-off of the Atlantis shuttle carrying the Galileo probe was delayed for several days because a small irregularity, which never recurred, was detected in one of many test cycles that were being run on a backup controller for the main engines.

In 1857, James Clerk Maxwell wrote a memorable letter to Michael Faraday, the first man to propose that magnetic fields existed. 'You are the first person in whom the idea of bodies acting [from] a distance by throwing the surrounding medium into a state of constraint has arisen,' Maxwell wrote, continuing: 'your lines of force can weave a web across the sky and lead the stars in their courses without any necessary immediate connection with the objects of their attraction.'

He was right.

Going Binary

Data are sent to and from spacecraft in the form of the coded binary digits (bits) used in computer systems. In a sense this is a kind of space-age Morse code, except that there are no 'dots and dashes' – simply the two states of pulse or no pulse.

The binary system is made up wholly of the symbols zero and one. A binary number has a given 'word' length with so many spaces which represent, from right to left, the decimal numbers 1, 2, 4, 8, 16, 32, and so forth. Thus the number 1 is 00001; 2 is 00010; 3 is 00011; 4 is 00100; 5 is 00101; 6 is 00110; 7 is 00111; 8 is 01000; 9 is 01001; 10 is 01010; and so on.

In actual telemetry operations, then, the sequence of time intervals silence-silence-silence-beep-silence in a five-bit word would represent the decimal number 2.

Rates of transmitting data are given in terms of how many bits are sent per second. Because of the variable distance along the line-of-sight path between the spacecraft and Earth, these rates must be adjusted periodically to make sure that good-quality information is received at Earth stations.

The early Mariner spacecraft to Mars transmitted data at the rate of 33.3 bits per second, meaning that it took more than 10 hours to send a photo from Mariner 4 to earth. But then computer and communication systems began to improve radically. Mariners 6 and 7 transmitted at 670 bits per second. With another huge spurt in technology, Mariner 9 sent information at 16,200 bits per second. Another technology leap, using X-band radio, brought Voyager to the 115,200 bits per second (115.2 kb/s) mark. This was Voyager's maximum transmission rate, achieved at Jupiter, its first planetary encounter. As distance increased the signal weakened and the transmission rate was adjusted to compensate – to 44.88 kb/s at Saturn and 21.6 kb/s at Uranus and Neptune.

By the time the Galileo system was built, technology had moved ahead to the point where data could be transmitted from Jupiter at 134 kb/s. With its high-gain antenna, Cassini is nominally able to transmit at rates from 5 bits per second to 249 kb/s.

Specialized coding methods make it possible to double bit rates without losing quality, and without adding to the tracking hardware. In the case of convolutional coding, the spacecraft telemetry equipment adds additional zeros and ones to the stream of bits in a predetermined rhythm. If the decoding computer at the tracking station 'knows' the coding scheme, it will detect and correct all bits that are in error. After decoding, the data is almost error-free.

Our Place in Space

'The investigation of nature is an infinite pasture-ground, where all may graze, and where the more bite, the longer the grass grows, the sweeter is its flavour, and the more it nourishes.'

T.H. Huxley

We know of no other planet that has an environment even remotely like ours. We are protected by an invisible magnetic bubble, enlivened by cycles of water and air, with the whole scenario perfected (from our perspective) by the forces of sunlight and the great heat source that lies deep below Earth's crust. Yet when we looked at Io's fountains we knew they were volcanoes, and when we saw the terraced hills of Mars we knew that lakes must have lain there. Our knowledge of the universe is guided by our observations of home, Planet Earth.

It's hard to determine when people finally decided that Earth was a planet, circling the Sun. No doubt the ancient Greeks suspected as much. The Church insisted that we had an Earth-centred universe. But Giordano Bruno (1548–1600) chose to differ. 'Innumerable suns exist; innumerable earths revolve around these suns in a manner similar to the way the seven planets revolve around our sun,' he wrote. 'Living beings inhabit these worlds.'

The brilliant, outspoken philosopher-poet, a refugee from holy orders, was burned at the stake because he taught that Earth circles its star. Galileo carried the torch carefully lest he meet the same fate, but for centuries now there has been no doubt that we're one of the solar nine.

Let us for a moment make a great perceptual leap, departing from ourselves to re-emerge in a small exploratory spaceship, just entering the human-inhabited solar system after that long flight from 51 Pegasi. We find a middle-sized, middle-aged star circled by countless bits of matter that were somehow left over from the original nebular omelette. We train our telescopes on the system. Of the circling bodies, nine globe-shaped specimens are of consequence. The innermost is what one would expect – since it's close to the star, virtually all the gas it contained in the past evaporated long ago and has

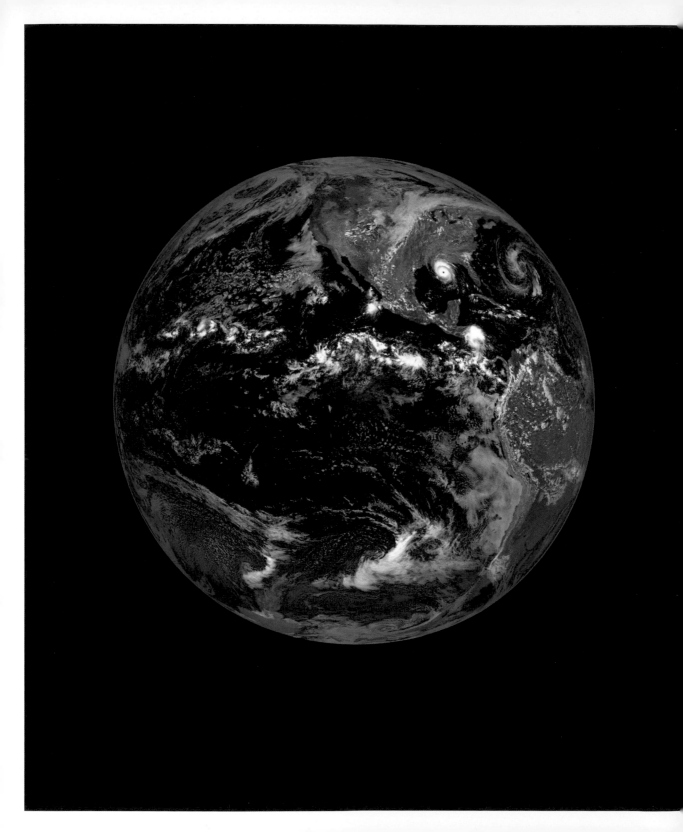

been unable to protect it from a vigorous peppering of meteorites that pockmark its surfaces with craters. Next to it is an inhospitably hot, lava-crusted orb cloaked in poisonous gases. Separated from the others by a thin ring of debris, four giant outer worlds still retain much of the gas that they had in the early part of their existence. The outermost planet, barely worthy of the name, is tiny by comparison and little more than an ice-ball.

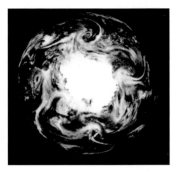

That leaves two. The first one encountered on the starward side of the debris ring is a reddish globe with a filmy atmosphere. The remaining one is a real oddball. It harbours many life-forms, and, wonder of wonders, its gaseous envelope is made up mostly of nitrogen and oxygen. Considering what we'd found on the other planets, hydrogen and helium would be likely components. But nitrogen and oxygen?

Its dominant life-forms call this planet Earth, a linguistic leftover from the word for soil or dirt. Its magnetic field protects it from the electrified blast of particles thrown out by its parent sun. Its gas blanket is so thick that it protects the surface from most meteoroids – they almost always collide with atmospheric particles and burn up from the friction before they hit the ground. So you don't see the remains of many impacts down there. What you do see is a surface that is two-thirds covered with water (the blue part), the remainder being exposed to the atmosphere (white for clouds and snow, and brown, with small spots of green). The surface is peculiar by comparison with other planets – great rocky plates float about like gargantuan ice floes, climbing under and over each other and making mountains and volcanoes in the process. Close up, mostly clustered near water masses, can be seen artificial constructs of vaguely geometric design, locally rich in atmospheric oxides, veined with pulsing arteries for the transport of energy, water, and life-forms. These, we learn, are cities, fringed with mosaics of fields and farmlands to feed their residents.

It is awesome, and a little uncomfortable, to consider that over six billion people call Planet Earth their home. We, and our living neighbours, from the smallest virus to the largest whale, helped to make Earth what she is, an unpredictable world of moving air, water, land, and teeming life. Even the white cliffs of Dover are animal-made, the crushed skeletal remains of countless tiny fossilized sea creatures.

There are immense amounts of water in the solar system, but, to the dismay of the astrobiologists, most of it is frozen solid. And it seems that liquid water and heat energy – whether from the Sun or some internal planetary source – is needed to support life. Earth, of course, is invested with all of the above.

Above: A unique but physically impossible view of the southern hemisphere and Antarctica. Any spacecraft directly over the Earth's south pole would see only half of this image – the rest would be covered by darkness! This view was created by mosaicing together several images taken by Galileo over a 24-hour period.

Left: Our blue-green planet. A view of the Pacific Ocean and the Americas. Hurricane Andrew can be seen in the Gulf of Mexico, off the coast of Louisiana.

Relatively primitive life-forms began to change the planet long before larger animals arrived on the scene. Somehow, in the methane-rich early aeons of our planet, life came to be – possibly heat-loving, sulphate-using microorganisms that lived in and around volcanic activity – 2.7 billion years ago, and perhaps as long as 3.5 billion years ago. Then it transformed into new shapes so numerous that we can never hope to map them all: some bound for extinction, some, from the fern to the lowly snail, becoming all but immortal in the marvel of their success.

Plant ancestors – early algae and cyanobacteria – charged the atmosphere with oxygen, paving the way for air-breathers like ourselves. They provided food for the herbivores that eventually would feed us, and the rich store of oil and gas that we would use to drive a technological civilization. Animals, in consuming plants and other animals, produced free carbon dioxide and nitrogen for the world of plants, along with methane to replace a minuscule amount of the supply that had blown into space in Earth's early aeons. The planet changed even more due to geological processes, including the restive dynamism of our planet's core, fluctuations in the amount of sunlight reaching Earth, and the weather that churns the atmosphere continuously to the chaotic orchestration of sea and sun.

Nothing like this happened on our neighbour planets – though Mars may have come part of the way, billions of years ago.

Humanity and Planet Earth

It is humankind that has generated the awesome capability of accelerating changes in our planetary environment. Much of civilized life calls for burning things – in the old days we needed fires for warmth, cooking, waste disposal, and smelting; today it is much the same, only more of it, with the addition of the vast number of new technologies that depend on burning coal, oil, and natural gas. These processes increase the amount of oxides in the atmosphere, the amount of carbon dioxide having shot up by more than ten per cent in the last thirty years of the twentieth century. Carbon dioxide has the ability to trap heat that is radiated from Earth's surface, producing a greenhouse effect. Since we are burning more fossil fuels, and chopping down the great forests that take in carbon dioxide, the atmosphere is warming and it is dubious that other, as-yet unidentified natural processes can balance the effect as they did following the meteor strikes and great volcanic events of the past. Some believe that a global warming of just a few degrees could, by the middle of this century, raise the sea level significantly – perhaps disastrously – and drive the practice of agriculture farther north.

A report released in 2001 by the United Nations-sponsored Intergovernmental Panel on Climate Change detailed how much humans had contributed to the increase in global warming in the previous half-century. Put simply, our increasing usage of modern technology was consistently injecting heat-trapping gases into the atmosphere. With the surge in greenhouse warming, temperatures could increase by as little as 1.4°C (2.5°F) to as much as 5.8°C (10.5°F) by the year 2100. A 5.8°C increase would be catastrophic. But of course nobody can make a reasonable guess as to what will happen because the computer models are imperfect and nobody knows how humanity will develop in the rest of this century.

Another effect of human activity is the depletion of the ozone layer high in the stratosphere. Ultraviolet (UV) light from the Sun breaks apart oxygen molecules high in the upper stratosphere, so that there is a relative abundance of free oxygen atoms. When these combine with normal oxygen molecules (O_2), ozone molecules (O_3) are produced. Ozone depletion is apparently caused by a complicated chemical reaction that results from the use of chlorofluorocarbons (CFCs) as refrigerants and aerosol propellants. Ozone absorbs much of the ultraviolet radiation before it enters the atmosphere. While small quantities of UV are essential to human health, helping us to generate vitamin D, large amounts can cause skin cancer, cataracts, immunosuppression, and other problems.

Industrial processes, and the natural effects of human habitation, lead to the pollution of air, water, and earth with toxic materials. Nature can clean some of it up, but not quickly enough to keep pace with population and material increase. Over the past few decades, humanity has begun to realize its responsibility for taking care of the planet. Earth satellites are helping us to become better stewards of our environment.

Today, by putting together information from various Earth observing programmes, investigators are gradually gaining a better understanding of Earth's natural processes. The study of clouds, water and energy cycles, oceans, atmospheric chemistry, land surface, water and ecosystems, glaciers and polar ice – and of the solid Earth – is making it simpler to gauge human effects on the environment and to predict natural hazards.

Donald Brownlee
Department of Astronomy
University of Washington

How Rare is the Earth?

Earth-rise as seen from the Moon by the Apollo 17 astronauts in December 1972.

Compared to its siblings, Earth is a planet with remarkable properties that have fostered its ability to retain complex life for extended periods of time. A basic but fortunate happenstance is that Earth lies within the solar system's 'habitable zone,' the not-too-close/not-too-far region where heating by sunlight allows oceans of water to neither freeze nor be lost to space.

Earth also has plate tectonics, a large Moon, a vigorous biosphere, and an atmosphere that contains free oxygen and thermostatic temperature control. It also seems to have a fortunate mix of appropriate amounts of water, carbon and nitrogen that allow Earth's complex systems to maintain long term conditions that are stable and habitable for complex life.

Earth is drastically different from its immediate neighbours, Mars and Venus. The surface of Venus is so hot it actually glows and Mars, while more Earth-like, is a too-cold planet with an ultra-thin unstable atmosphere. Life surely cannot exist anywhere on Venus because even the fundamental molecules of life cannot survive the heat. Life might exist under the surface of Mars, warmed by geothermal heat and protected from the harsh sunlight. But, at best, Martian life is probably limited to microbial organisms analogous to the bacteria and archaeans that have dominated most of Earth's history.

Earth, as we know it, is rare not only in the solar system but even when it is compared with itself. Earth will last for twelve billion years before it is engulfed by the red giant Sun, but the tenure of animals and plants will last only a billion years. For its first two billion years of life, Earth had essentially no free oxygen in the atmosphere and for its first four billion years, life was invisible – microscopic. A sobering reality of our views of life in the universe is the fact that, on what seems to be a nearly ideal planet, it took four billion years of biological and planetary evolution just to get to animals.

Things will dramatically change in the future as the Sun becomes ever brighter with age. Within a half to one billion years into the future, plants and animals will no longer be able to live in the environments provided by Earth. On longer timescales, Earth will lose its oceans to space and the formerly blue planet will become a pink body coloured by oxidized iron and salt. An alien visitor looking at Earth randomly throughout its history would only see what we presently consider to be truly Earth-like conditions about 10 per cent of the

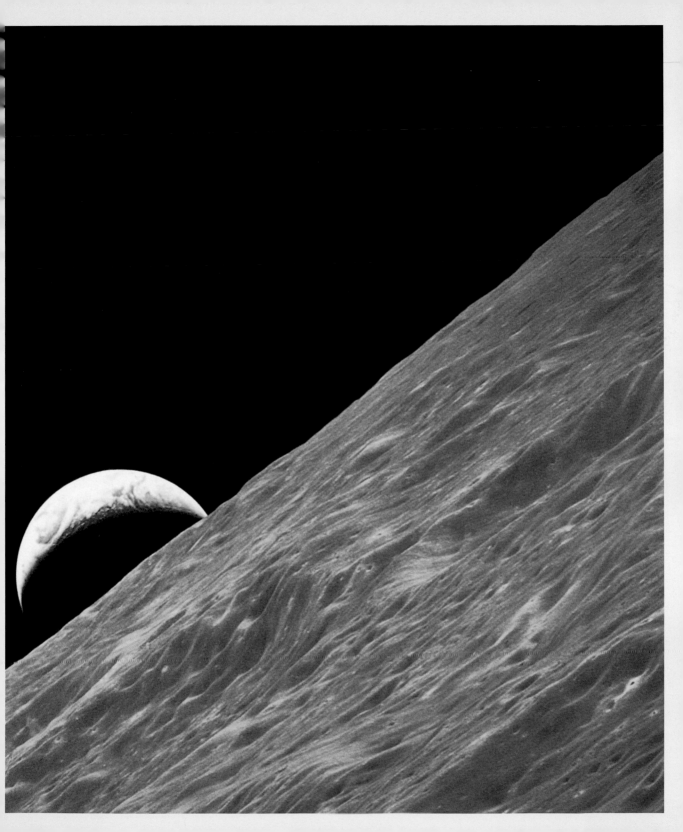

time. The same visitor randomly visiting all of the solar system planets over a ten-billion-year time-frame would find an animal-supporting life planet only about 1 per cent of the time.

Earth is rare in our solar system – but what about Earth-like bodies in other planetary systems? We know that at least several per cent of nearby stars have giant planets, and it is possible that the majority of stars harbour planetary systems. What fraction of stars have truly Earth-like planets in their habitable zones? Most of the known stars with planets (SWPs) probably don't have such planets because their giant Jupiter-like planets have small or elliptical orbits that would cause instability problems for truly Earth-like planets. In most of the known SWP cases it is likely that the motion of the giant planets would have destroyed Earth-like planets on circular orbits.

The ultimate question of the frequency of Earth-like planets depends on how Earth-like a planet really has to be to harbour Earth-like life. Odd factors such as the presence of Jupiter, asteroids, and comets, our location in the Galaxy, our time of formation in the Galaxy, the metallicity and mass of the Sun, the mass of our own planet, and a host of other influences were involved in making our planet what it is.

There is also the role of chance. The more we learn about planets and planetary systems, the more we realise how complex they can be and how diverse their evolutionary pathways can be. We will only know for sure 'how rare Earth is' when we detect other terrestrial planets around other stars and have the ability to compare them with our home. If true Earth-like planets with oceans and biospheres are common, then it should be possible to find them orbiting nearby stars. If they are rare, we will probably never know just how rare they are. If they are too rare, then the nearest ones will be too far away to study and we will be effectively and forever isolated from them by the vast distances of interstellar space.

It is nearly inconceivable that we could ever be truly 'alone' but if we are rare enough we can effectively be alone in the sense that we could be essentially isolated by space and time.

The First Map-Makers

Things were simpler for our early ancestors. Science and technology were the protected provinces of philosophers, warlords, and clergy. Earth was a flat place bounded by geographical features such as coastlines and mountains. 'Mediterranean' means 'centre of the Earth' – and so it was in Homer's time, when the world was thought to be surrounded by a single great sea. The idea that one could fall off the edge of the Earth, or sail over it, lingered for a remarkably long time, to be disproven entirely when the great explorers finally circumnavigated the globe in the sixteenth century.

But it happens that, in the city of Alexandria, in the third century BC, there lived a thoughtful librarian named Eratosthenes. A few decades earlier, Euclid had founded the science of geometry in that same great centre of ancient learning. Among his books Eratosthenes read a notation that something strange happened every June 21 at a town named Syene, up the Nile and not very far from the Equator. If one stuck a straight stick into the ground at noon on that day, it would cast absolutely no shadow. Eratosthenes was a scientist as well as a bibliophile, and his curiosity was piqued. He placed a stick in the ground back home in Alexandria, observed it at noon the next time June 21 rolled around and it did cast a shadow. How could that be? More to the point, how could that be if the world were flat and the Sun directly overhead? To solve the problem, Eratosthenes had to assume that the Earth was round. Then he took his reasoning a little further. He knew the distance from Alexandria to Syene was 800 kilometres (500 miles), and knew that, if he extended the length of those two sticks far enough that they would touch, the angle of divergence would be seven degrees. Euclid's geometry also told him that if Earth's circumference were circular, it would be 360 degrees around. Since seven degrees is about one-fiftieth of 360 degrees, fifty times 800 kilometres – 40,000 kilometres (roughly 25,000 miles) – should be the circumference of the planet. He was right on target.

Eratosthenes' work was not universally accepted, even though sailors could quite clearly observe the gradual disappearance of ships as they sailed into the horizon – hull first, finally the topmasts – and everyone could see the curved shadow cast by the Earth on the Moon during eclipses.

Naturally it didn't take an Eratosthenes to draw a map. A tablet from about 1000 BC, discovered in Iraq, depicts the Earth as a disc surrounded by water, centred by the city of Babylon. Maps had a highly utilitarian purpose until the Greek intellectuals came along: mostly, they existed as a way of indicating political or trading boundaries. The first geography book was produced in about 500 BC, scribed by Hecataeus and portraying Earth as a disc. Pythagoras (sixth century BC) and Parmenides (fifth century BC) liked the idea of a spherical Earth and, after Aristotle championed their cause in about 350 BC, the idea became generally accepted among scholars .

Ptolemy, born Claudius Ptolemaeus (c. AD 90–168), was the great man of ancient geography and cartography. A student at the Alexandrian library, he produced an eight-volume guide to geography, complete with a guide to mapmaking and a list of about eight thousand places, their latitudes and longitudes. Ptolemy's scholarship was profound, but he underestimated the size of the world – a factor that, more than a thousand years later, caused Columbus to underestimate the distances to Cathay and India.

At least partly due to religious reasons, the *Geography* – carefully protected by Arab scholars – was not translated into Latin until about 1405, though there exists a remarkable map of Europe that was drawn by Catalan cartographers in 1374. Then the western Europeans finally got into the act. Their age of discovery was just beginning; they had improved instruments, including the compass; and they had improved ships with which to ply the oceans. The cross-staff, the sextant, and the theodolite were used to determine latitude. Soon Ptolemy's map was being updated with data brought home by the likes of Columbus, Cabot, Magellan, da Gama, and Vespucci. The most famous mapmaker of the time was the Flemish scholar Gerhard Kremer, or Gerardus Mercator (1512–1594). Mercator devised a projection that carries his name, which made it possible for navigators to plot bearings as straight lines, and published a complete map of Europe in 1554. North and South America, as continents separate from Asia, were first depicted in a huge map produced in 1507 by Martin Waldseemüller (c.1475–c.1520). Impressed by the writings of Amerigo Vespucci, Waldseemüller was probably the first to put the name 'America' in print.

The next surge of improvements in mapmaking occurred in the 1700s, with new demands brought by territorial expansion and the need to improve civil administration. The accurate measurement of longitude was finally made possible with the introduction of John Harrison's chronometer, which, since it didn't need a pendulum, made it possible to take Greenwich time reliably to sea. The idea was to compare local solar time with universal time, then calculate exactly how many degrees of longitude were represented by the time difference. The breakthrough occurred after Harrison (1693–1776), a Yorkshire carpenter, entered the British Government's contest to find a way of

The volcanoes Kliuchevskoi, Bezymianny and Tolbachik, Kamchatka, Russia. This image was acquired by the Spaceborne Imaging Radar-C and X-band Synthetic Aperture Radar (SIR-C/X-SAR) aboard the Space Shuttle *Endeavour*. SIR-C/X-SAR illuminates Earth with microwaves, allowing detailed observations at any time, regardless of weather or sunlight conditions. SIR-C/X-SAR imagery typically exaggerates the height of its vertical features.

The Kamchatka volcanoes are among the most active in the world. The volcanic zone sits above a tectonic plate boundary, where the Pacific plate is sinking beneath the northeast edge of the Eurasian plate.

determining a ship's longitude within 30 miles (about 48 kilometres) after a six-week voyage. He passed the test in 1735 – but had to wait until 1773 to get the final instalment of his prize money.

The Earth's mass was determined to be 5.98 x 10^{21} tonnes by England's Charles V. Boys (1855–1944) and Karl Ferdinand Braun (1850–1918) of Germany. To do this, they made use of Newton's law of universal gravitation, which had been used by Henry Cavendish in 1797–98 in experiments that determined the gravitational constant. Cavendish used two pairs of lead spheres to reach his conclusions. In separate experiments Boys and Braun compared the gravitational attraction of a small sphere at the Earth's surface with that of a large sphere of known mass on the same small sphere. After the volume and mass were known, all it took was simple mathematics to calculate the mean density of Earth – 5.527 grams per cubic centimetre (338 lb per cubic foot), which by definition is 5.527 times the density of water. It is remarkable that such a simple experiment could yield such accurate results. Today's gravimeters work in much the same way, measuring gravitational attraction to an accuracy of within one part in ten million by observing how much a small mass is deflected by a large mass. These refined instruments have helped to redefine Earth's mass as 5.9742 x 10^{21} tonnes, and revise its density downward to 5.515 g/cm^3 (337 lb per cubic foot).

Private mapmakers eventually were supplanted by military establishments, and these by civilian institutions including the Ordnance Survey of Great Britain and the Institut Géographique National of France. Other industrialized nations formed similar organizations in the nineteenth century, including the Geological Survey and National Ocean Service in the U.S.

Earth Satellites

By the end of the Victorian era we had learned the shape of the Earth in three dimensions – a job that we had done extremely well with experiments performed on the surface of the Earth. We were now about to embark upon a new phase in which we would explore more about what makes Earth tick – its environment and the dynamics of that environment. For this, we would have to observe the Earth from afar.

In the twentieth century, wartime again increased military involvement and introduced aerial surveys. Then, with the advent of space technology, satellite imagery added new detail, particularly in the case of mapping features that are difficult to measure at close quarters. These included the physical characteristics and movements of air, water, and even earth.

In 1945, Project RAND (now RAND Corporation) issued a secret report that recommended U.S. development of 'a satellite vehicle with appropriate instrumentation' that might well prove to be 'one of the most potent scientific tools of the twentieth century.' In 1948, the U.S. Department of Defense announced that an Earth-satellite programme had started. In 1953, a satellite design was announced by an Anglo-American team that included Arthur C. Clarke and Fred Singer of the U.S. Office of Naval Research. Their proposal was MOUSE – 'Minimum Orbital Unmanned Satellite of the Earth' – a battery-powered package containing a transmitter and instruments for measuring solar ultraviolet radiation, gamma rays, and solar x-rays. It never flew, but it made people think. And it came at a good time, just before the International Geophysical Year, which began in 1957–58.

In 1958, hard on the heels of the Soviet propaganda success with Sputnik, the first successful scientific satellite, Explorer 1, was launched by the U.S. It discovered and measured the first of the two Van Allen belts of charged particles that circle the globe. Early satellites also showed that the Sun causes the atmosphere to swell by day and contract by night, mapped the exact shape of the slightly flattened Earth – its geoid – for the first time, and continuously monitored the movement of the drifting continents.

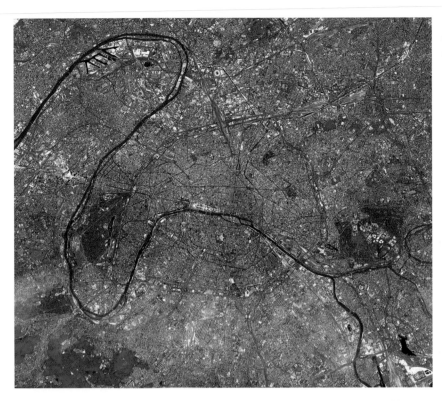

Left: This view of Paris was created by the Advanced Spaceborne Thermal Emission and Reflection Radiometer (ASTER) instrument on NASA's Terra satellite. ASTER was designed to monitor and map the changing surface of our planet.

Below, left: Monitoring glacial retreat, this ASTER image covers an area of 55 by 40 kilometres (34 by 25 miles) over the southwest part of the Malaspina Glacier in Alaska. Here, the ice and snow are shown in light blue.

Below, right: An ASTER image of northern Shanxi Province, China, highlights a section of the Great Wall, visible as a black line running diagonally through the image from lower left to upper right. The Great Wall is commonly cited as the only man-made object visible to the naked eye from space.

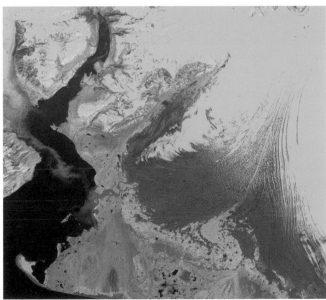

Today, thousands of Earth satellites cross the skies, sponsored by more than a dozen different countries. On a day to day basis, they map the Earth in marvellous detail. The Stanford Radar Interferometry Group alone reveals centimetre-size changes in the Earth's crust. These might include land subsidence near water-thirsty Las Vegas, glacial movement in Greenland, earthquake deformations in California, and tides caused by the Moon in Earth's surface as well as in its seas.

Two satellite launches – the U.S. Landsat-5 in 1984 and the French SPOT-1 in 1986 – started a new age of large-scale terrestrial mapmaking. The science had already moved from ground-based observations to aerial mapping; now began the era of Earth-satellite imagery.

Today, remote imaging is a business, with high-resolution images capable of showing temperature and elevation as well as outlines and colours, for sale to farmers, fishermen, oil/gas exploration companies, and town planners. The computer sciences are extremely important partners, making it possible to enhance the original images for specialized purposes, and when desired to overlay information such as roads, rivers, and power lines.

Landsat-5 caused a sensation by providing colour photos in which it was possible to distinguish features as small as 30 metres (100 ft). In 1992 Russia offered 5-metre (16.5-ft) resolution. Then, in 1999, Space Imaging, a private company, brought the resolution to 1 metre (40 in) with its IKONOS satellite system. At the same time, synthetic aperture radar (SAR), offered by the Canadian RADARSAT-2, was offering black and white images with 3-metre (10-ft) resolution – at night and through cloud if necessary.

SAR is responsible for some of the most exciting images of Earth and other planets published in recent years, in terms of visual appeal as well as data content. To make a synthetic aperture radar image, about 1500 high-power radio pulses per second are transmitted toward the imaging area, each pulse lasting between 10 and 50 microseconds. Part of the energy in the radar pulse is reflected back toward the antenna, which picks up the returning echoes. These echoes then are digitized and stored for processing and display. Because the satellite is moving at high speed, it picks up different parts of the returning echo from different positions, and the data can be treated as if the satellite were tracing a line across a much larger antenna (hence the term synthetic aperture), with a corresponding increase in resolution.

Today, the images we see of Planet Earth are seldom ordinary photographs. They are representations of data, reproduced by computers in the form of images. Not infrequently, they are combined with other data-images to provide a picture that is relevant to our specific interests. In 1989, 'The Breathing Earth,' touted as the first image of the global biosphere, was produced by combining NASA data from the Nimbus-7 remote sensing satellite with data from a U.S. National Oceanographic and Atmospheric Administration satellite NOAA-7. An image of the huge Chicxulub impact crater in Mexico was made using gravity and magnetic field measurements that were able to penetrate layers of sediment. In 2000, an impressive view of California's famous San Andreas Fault consisted of radar images made from the Space Shuttle Endeavour, combined with a colour Landsat satellite image.

The first major satellite mapping project of the twenty-first century went into service in March 2002. This was the European Space Agency's Envisat, carrying instruments from many nations in a payload integrated at the Matra Marconi Space Laboratory (by then absorbed into the Astrium organization) in Bristol, England. Specialized for climate research, it includes synthetic aperture radar; a spectrometer for colour studies of the ocean, atmosphere, and land; another spectrometer to measure the gaseous makeup of the atmosphere; a specialized sensor for measuring ozone concentrations; and a radar altimeter to study the roles of oceans and ice in global climate.

Envisat is a large satellite that towers several storeys high. But the profusion of map-making satellites – and especially of communications satellites – is

Spaceborne radar imaging reveals the scars left by the impact of an asteroid or comet several hundred million years ago in the Sahara Desert. The Aorounga impact crater in northern Chad has a diameter of about 17 kilometres (10.5 miles) and is possibly just one of a string of impact craters formed by multiple impacts. Radar imaging is a valuable tool for the study of desert regions because the radar waves can penetrate thin layers of dry sand to reveal structures that are invisible to other sensors.

being spurred by the development of small payloads, commonly known as nanosatellites, which make it possible to use smaller and cheaper launch rockets. The U.K.'s Surrey Satellite Technology, a pioneer in this field, produced the 6.5 kg (14.3 lb) SNAP-1, launched in 2000 on the same Russian Cosmos rocket as a Chinese Tsinghua disaster monitoring satellite (also built by SST). SNAP-1, the first nanosatellite with a propulsion system to perform orbit-changing engine firings, rendezvoused with the Tsinghua satellite in orbit using the U.S. Global Positioning System (GPS) for orbital navigation, and delivered images back to its operators on Earth.

The world's largest collection of satellite images and aerial photographs of Earth's land mass is held by the U.S. Geological Survey's (USGS) National Mapping Division, at its Earth Resources Observation Systems (EROS) Data Center in the farmlands of Iowa. Since 1991, the Eros Data Center has supported the United Nations Environment Programme/Global Resources Information Database, making environmental data available to developing countries. Many artificial Earth satellites – apart from communications and military satellites – have missions that are intended to predict natural hazards and monitor human effects on the environment. In the U.S., much of this work comes under the aegis of NASA's Earth Science Enterprise, formerly Mission to Planet Earth; socio-economic and Earth science data are combined at the agency's Socio-economic Data and Applications Center (SEDAC).

The Terrestrial Sphere

Thanks to the work of scientists using ground- and space-based equipment, with a continuously improving armamentarium of instruments and computers, we know Earth in intimate detail. It is a place with oceans that cover more than two-thirds of its surface; with continents that float almost immeasurably slowly on a deep sea of molten rock; with a molten nickel-iron core that, as it spins, generates a magnetic field fashioned into a teardrop shape by the solar wind of charged particles from the Sun; and where some of those same particles are caught up in the field and transported to the poles, producing brilliant aurorae. Oddly, and for reasons that nobody yet knows, the magnetic field sometimes flips, and the north pole becomes the south.

Overall, more than 90 per cent of our planet's mass, including core, mantle, and crust, is represented by four elements – iron, oxygen, silicon, and magnesium. About one third of the mass is represented by the hellish, radioactive combination of molten iron and nickel concentrated in the core. Its temperatures are incredibly hot, approaching or surpassing that of the Sun's surface as they rise to more than 7,000°C (12,600°F) at the Earth's centre. Encircling the core is a thick, silicate-rich mantle that reaches upward nearly 3,000 kilometres (1900 miles) to the planet's crust. This outermost layer consists of more than a dozen separate rigid blocks, or plates, that can be as thick as fifty kilometres (30 miles) below the continents, but perhaps only one-tenth that amount below the ocean floor.

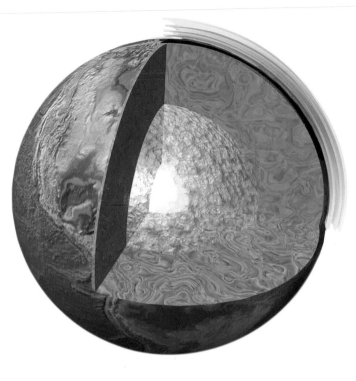

The Earth, pictured as a series of spheres, one inside the other, each one hotter and denser than the previous one. The outermost sphere is the atmosphere which extends approximately 400 kilometres (250 miles) into space. This rests on an average of 80 kilometres (50 miles) of lithosphere made up of crust and rigid upper mantle. The lithosphere 'floats' on the semi-plastic asthenosphere, beneath which lies the lower mantle. At a depth of nearly 3,000 kilometres (1,900 miles) lies the core. The outer core is a white-hot liquid composed mostly of iron, nickel and sulphur. The inner core is a solid sphere of iron-nickel alloy hotter than the surface of the Sun and under immense pressure. Heat from inside the Earth keeps the planet dynamic, fueling volcanoes and earthquakes.

Our knowledge of the regions below Earth's surface owes much to the practice of seismic mapping – the use of artificial earthquakes, often to find underground strata that might contain natural gas or oil. When an explosion or mechanical 'thump' is triggered at one location, sound waves travel downward, then bounce back from subsurface layers to microphones at another location. As this process is repeated, a map is constructed that indicates the shape and consistency of the subsurface layers.

Seismology's greatest gift to geology occurred in 1909, when a Croatian scientist, Andrija Mohorovicic (1857–1936), discovered that waves were being reflected from a depth of about 60 kilometres (38 miles). This turned out to be the bottom-most layer – the boundary between the crust and the mantle. Mohorovicic's experiment was repeated over and over and again, all over the world, with similar results.

Then, in 1915, after noticing how the various continents can be made almost to fit together in the map of the world, the German geologist Alfred Wegener (1880–1930) published his theory of 'continental drift.' Wegener theorized that, two hundred million years ago, there was a single land mass, which he called Pangea, which then fractured and formed different continents. Now we know that the Earth's crustal plates move around on a slightly fluid layer (the aesthenosphere) in the process called plate tectonics. Pangea was formed 350 million years ago by the coalescence of at least three major continents – Gondwana (or Gondwanaland), Laurentia, and Baltica – and began to break up about 100 million years later.

Due to tectonics, the North American continent moves westward on its gigantic plate at the rate of 2–3 cm (roughly 1–1.5 in) per year, engorging layers of rock beneath its west coast and processing them in a natural furnace of enormous heat and pressure. Into Earth's maw is drawn limestone that enriches volcanic plumes with carbon dioxide (which eventually ends up locked away in limestone again as the process continues), and vegetation that is converted into deposits of coal and gas.

Earth's volcanism can provide insights into the nature of phenomena that occur on other planets. 'Extraterrestrial' analogues on Earth's surface include the maw of Costa Rica's Poas Volcano. Hardly a holiday retreat, the place bubbles with waterless ponds of liquid sulphur – miniature cousins to the huge bodies of sulphur that dot the surface of Jupiter's turbulent volcanic moon, Io.

It's estimated that more than a hundred tonnes of extraterrestrial material lands on Earth every year, small chunks of rock that have somehow survived entry into the atmosphere. During its first billion years, Earth was struck by billions of icy bodies, a process that probably invested it with its supply of water, which now accounts for 0.1 per cent of the planet's mass.

As for truly large impacts, about 150 craters have been discovered – a modest number by comparison with other planets. They include the Barringer meteor crater, a kilometre-wide (0.6-mile) landmark in the barren desert of eastern Arizona, as well as the Chicxulub crater in the Gulf of Mexico. Meteor Crater, estimated to be 49,000 years old, was the first on Earth to be identified as an impact crater, in 1920.

Cloaking the solid sphere of Earth (the lithosphere) are the hydrosphere and atmosphere, sometimes together called the ecosphere, and the magnetosphere.

The hydrosphere consists of all the Earth's water – free, frozen, and vaporous, fresh and salt – above the ground and below. About 13 per cent of it is ice; less than 1 per cent is represented by rivers, streams, and lakes. All is part of the life-giving water cycle, where water evaporates, forms clouds, falls as rain or snow, then evaporates again from sea or land, starting the process anew. NASA's SEASAT satellite, which briefly flew a polar orbit in 1978, first proved the capabilities of radar for determining wave heights, the amount of water vapour in the atmosphere, and the roughness of the ocean surface.

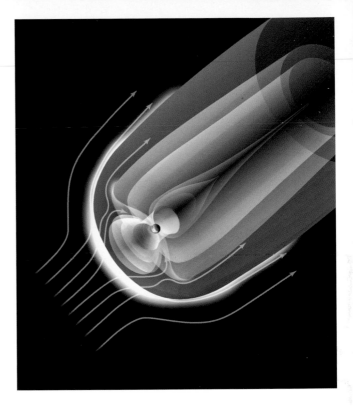

As the Earth rotates, the liquid outer core spins, creating the Earth's magnetic field. The magnetosphere is shaped by the interaction of the Earth's magnetic field with the solar wind.

Though it is 100 kilometres (62 miles) thick or more, the atmosphere is a thin shell by comparison with the solid Earth. The troposphere, where life and weather exist, is even thinner, measuring just 12 kilometres (7.5 miles) thick. It consists of 78 per cent nitrogen and 21 per cent oxygen. Above the troposphere is the stratosphere, a dry region almost devoid of wind that reaches to about 50 kilometres (31 miles) altitude. From that point, the atmosphere gradually gives way to a region characterized by a relatively large number of charged particles – the ionosphere – and dissipates in the magnetosphere.

The magnetosphere, shaped by the interaction of Earth's magnetic field with the solar wind, extends to much as 60,000 kilometres (37,500 miles) toward the Sun, where the oncoming stream of particles meets it and forces it back in an elongated 'magnetotail' that stretches several million kilometres into space. NASA's Magsat, which flew from 1979 to 1980, was one of the first spacecraft to acquire accurate data on the magnetic field. The European Space Agency's Cluster satellites – Salsa, Samba, Rumba, and Tango – flew in formation through the magnetosphere in 2000–2002, exploring its interaction with the charged particles swept along in the solar wind.

For all this research, we still do not know what makes Earth's magnetic fields 'flip' every million years or so; we do not know why the atmosphere contains so much of a highly reactive gas – oxygen – that normally would combine with other chemicals; we do not know how weather is changed by the interaction of solar energy with high-altitude clouds. Despite all the discussion, we do not even know the true mechanisms of global warming or the growing ozone hole.

Fortunately, our experiences in exploring Earth have given us the tools to find out, and the insight to foresee, what might be awaiting us on the other planets and their large moons. Even before the advent of space technology, we used latitude and longitude in our planetary mapmaking, and put together gradually more detailed images of the surfaces of other worlds through telescopic observations. This was visible information. We knew the distance of those faraway bodies, their size, mass, and kinetics, even something of the surface features – the Moon's craters, the Great Red Spot of Jupiter, the seasonal colour changes of Mars. But we wanted to know more. With our imaginations fired by exciting discoveries about Earth, we wondered whether the other planets had magnetic fields and aurorae, tectonic activity, underground water, and greenhouse effects. Did they have analogues of Earth's hydrosphere, atmosphere, and magnetosphere? The terrestrial planets certainly had solid rock and sand surfaces, but what about the giant outer planets? What lay beneath their thick cloud decks? Only spacecraft could answer these questions.

Earth is our home, our parent and our grave combined; for all intents and purposes it is indivisible from us. The human capacity for thought also enables us to appreciate and stand in awe of Earth, and the other planets, stars, and galaxies beyond. Cicero (106–43 BC), would be just as quotable if he were a modern-day Nobel Prize winner: 'The contemplation of celestial things will make a man both speak and think more sublimely and magnificently when he descends to human affairs.' And Galileo, turning from his primitive telescope, reflected that 'philosophy is written in this grand book – I mean the universe – which stands continually open to our gaze.'

It is no secret that our lives are patterned by personal experience, and by our observations of the experience of others. So it is with the entire human family, staring out from a place that we know to others that we do not. Our knowledge of home helps us to understand the ways of planets beyond our ken.

Earth's Moon

Sometime in the distant past, when the planets were just being formed, the rather plastic mass that was destined to become Earth suffered an unimaginable catastrophe. From somewhere within the roiling disc of gas and dust that surrounded our young star, there emerged a huge chunk of rock that might, had conditions been different, have become a planet itself. It was on an unchangeable collision course with the future Earth.

Our planet was partly formed at the time of the collision, so that a core and mantle were already in place. The collision that obliterated the invader completely also blasted off a very large amount of silicaceous mantle material from our planet, accompanied by a relatively small amount of molten iron from the core. As the fragments of rock coalesced together again in orbit around our wounded planet, they formed our satellite.

The Moon is a peculiar satellite by comparison with similar bodies around the solar system. It is relatively large, so that in some respects the Earth-Moon relationship can be considered to be that of twin planets, like Pluto and Charon at the edge of the solar system. It is slightly egg-shaped, with the 'point' aimed at Earth. Its gravitational pull causes large tides in its parent planet's oceans and minute tides in its crust. In turn, it experiences Earth-generated tidal influences that probably cause moonquakes and the release of small amounts of gas from the interior.

The Moon has a diameter of 3,476 kilometres (2,160 miles) and revolves around our planet at an average distance of about 384,400 kilometres (238,900 miles). While it is nearly a quarter the size of Earth, it weighs just one-eightieth as much, with a density about 3.34 times that of water. Lunar days approximate the lunar cycle we know on Earth – the satellite circles us once every 27.32 days. Because there is little or no atmosphere, temperatures are extreme. They range from 107°C (225°F) in the day to -153° C (-243°F) in the lunar night. As is readily apparent, the Moon has many craters, and craters within craters, sometimes overlapping, that resulted from high-speed meteorite impacts. These range up to 2100 kilometres (1300 miles) in diameter, though most are far smaller. The surface also has dark areas known as maria or seas, marking where lava once flooded the major impact basins, along with mountain ranges and deep parallel trenches (rilles) that may be hundreds of kilometres in length.

Naturally the Moon has been observed, and sometimes worshipped, since the early days of humankind. Because it is so prominent in our sky, so that we can see the craters with the naked eye, it has been a popular object of fantasy

On its way to Jupiter, the Galileo probe turned its sensors towards the Moon. This image shows compositional variations in parts of the Moon's northern hemisphere. Bright pinkish areas are highlands materials, such as those surrounding the oval lava-filled Crisium impact basin towards the bottom of the picture. Blue to orange shades indicate volcanic lava flows. To the left of Crisium, the dark blue Mare Tranquillitatis is richer in titanium than the green and orange maria above it. Thin mineral-rich soils associated with relatively recent impacts are represented by light blue colours; the youngest craters have prominent blue rays extending out from them.

and speculation. Giovanni Riccioli (1598–1671), an imaginative seventeenth century Italian Jesuit, named two of the blue 'seas' the Sea of Showers (Mare Imbrium) and the Sea of Nectar (Mare Nectaris). H. G. Wells (1866–1946) wrote of an underground lunar civilization in his *First Men in the Moon*.

The mystery of why the Moon appears larger at the horizon was explained by Ptolemy – things simply seem bigger when they are part of 'filled space,' such as a skyline. As for the blue Moon commemorated by the old song: the Moon looks blue only when it is seen through particles that will scatter its light the right way, like the smoke from forest fire – an occasion so rare that astronomers use the term informally to describe the full Moon when it appears for a second time within a single calendar month.

The Moon lost much of its romance as scientific truths came to light during the U.S. Apollo programme. Fuelled largely by political motives – the U.S. needed inspiration during a gloomy time in the Cold War – Apollo produced eleven crewed lunar missions, six of which actually landed, including the first Moon landing (Apollo 11) on July 20, 1969. It yielded considerable data on the geology and geophysics of the Moon – and many photographs, including the incongruous videos of bulky-suited astronauts bunny-hopping over the lunar surface. Perhaps just as important was the experience gained in human reactions to spaceflight, extravehicular activities, and complicated manoeuvres such as docking. The programme was only marred by a launch pad fire in January 1967 that killed the three-man crew of Apollo 1.

Robot programmes yielded, and continue to yield, a rich harvest of lunar information. Russia's Luna 3 photographed the far side of the Moon in 1959, the United States' Project Ranger provided the first close-up images all around the Moon in 1964–5. Luna 16 landed in September 1970, took a sample of lunar soil and brought it back to Earth. A year later, Luna 17 took a highly successful, TV-equipped robot roving vehicle, Lunokhod I, to the Moon's surface. NASA's Project Surveyor, meanwhile, provided images from the Moon's surface and analyses of the chemical composition and mechanical properties of the Moon's soil. By the time the first astronauts arrived, the Moon had been mapped in detail by America's Lunar Orbiter missions.

The Galileo spacecraft, picking up speed before its journey to Jupiter, imaged the Moon during brief flybys in 1990 and 1992. Then the Clementine spacecraft obtained detailed images and mapped the topography of the Moon from orbit in 1994. At the turn of the twenty-first century, exploration of the Moon was dominated by the results of data reduction from NASA's Lunar Prospector, which went into orbit in January 1998. In March of that year,

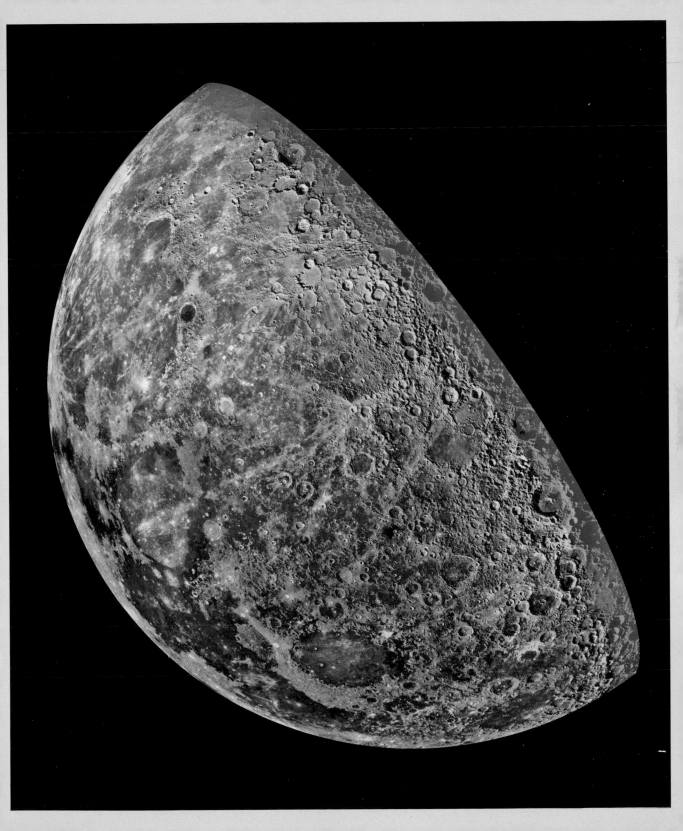

A mosaic constructed from approximately 1,500 Clementine images of the south polar region of the moon. The polar regions of the moon are of special interest to scientists because of the potential occurrence of ice buried in permanently shadowed areas such as deep craters. The south pole is of greater interest because the area that remains in shadow is much larger than the corresponding area at the north pole.

mission scientists exuberantly announced that they might have detected water ice in craters near the poles. The ice, accumulated in areas forever shadowed from the Sun, could conceivably be the 'tip of the iceberg,' the visible portion of up to six billion tonnes that lay buried in the regolith.

This was extremely exciting information – but was it true? The data came from neutron spectroscopy, which indicates the presence of water ice by measuring how much neutrons are slowed down by hydrogen ions. Working together, engineers from NASA and the University of Texas then decided they needed to make a spectrographic analysis of the material in those tantalizing polar craters. They decided to deliberately crash the spacecraft into the south pole, and see if ultraviolet emission lines showed up in spectrometer observations of the crash site. Ultraviolet indicates the presence of the hydroxyl chemical group, and this is what the experts call a 'water signature.'

The crash occurred on schedule, but the results were negative. The odds in favour of lunar water dropped. But it could have been a fluke. Scientists came up with eight reasons why this might have been the case, from the spacecraft hitting a rock to the telescopes pointing in the wrong direction.

Human exploration and colonization of the Moon lost popularity toward the end of the twentieth century, when Mars was adopted as a more attractive target. A NASA report, 'Leadership and America's Future in Space,' issued in 1987, recommended that a lunar base be built in three phases – robotic exploration; construction of a small, six-astronaut base; and then enlargement of the base and the initiation of large-scale scientific research. It was then suggested that the Moon's human population might reach 100 by 2010, but by the beginning of the new century the report was all but forgotten.

Meanwhile, in Japan, engineers were putting together SELENE (SELenological and ENgineering Explorer), the country's first major lunar probe. The National Space Development Agency of Japan (NASDA) is interested in scouting the Moon's terrain for future exploration and exploitation, as well as to gain scientific data on the Moon's origin and development. Their 'Scenario of Lunar Base Construction' envisages another three-phase programme, to be completed in 2024. It would consist of reconnaissance and robotic exploration, robotic construction of a base, and human occupation.

In Europe, ESA's SMART-1 mission to the Moon (scheduled for a 2003 launch), is designed as a geophysical explorer. But there is more. 'We are in an exciting phase of international exploration,' said Project Scientist Bernard H. Foing, 'starting now with orbiter and penetrator precursor missions, to be followed by landers, rovers, lunar robotic villages... This prepares the possibility for permanent human presence on the Moon and in the Solar System.'

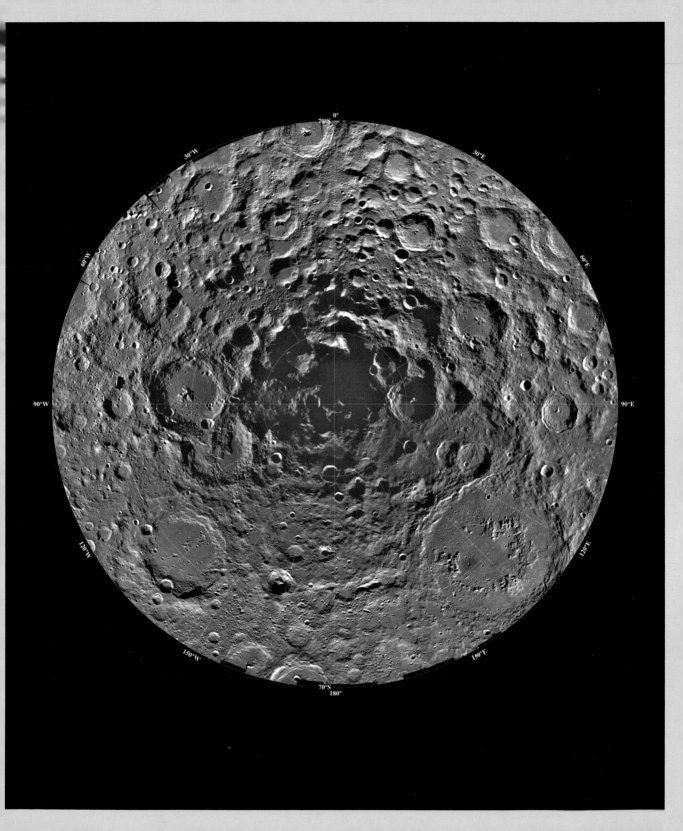

The Planetary Neighbourhood

'What now, dear reader, shall we make of our telescope? Shall we make a Mercury's magic wand to cross the liquid aether with, and like Lucian lead a colony to the uninhabited evening star, allured by the sweetness of the place?.'

Johannes Kepler

Observations of Earth provide clues to the character of our planetary neighbours – Mercury, Venus, and Mars. However, comparisons of the four show that, while all were derived from the same region of the original solar nebula, nothing much remains of a family resemblance. Each has its own long and distinctive history, shaped by many factors, including distance from the Sun, cataclysmic collisions with other orbiting bodies, and even proximity to the huge gravitational field of Jupiter. From each is unfolding a fascinating story – from moon-like Mercury, from cloud-shrouded Venus, and from what has proven to be the most attractive of them all, Mars.

Earth's companions in the inner belt of planets are Mercury, Venus, and Mars, in order of closeness to the Sun. In the early days of the solar system, they emerged with Earth from the chaotic violence of clashing material and gas that was their cradle, in the close precincts of the Sun. Policed and shepherded by the enormous hulk of the adolescent Jupiter, they were gouged, sliced, pocked, and bulldozed by everything from meteoroids to other would-be worlds. The marks of violence remain today.

These were the earliest known planets. Mercury, which zips around the Sun every 88 Earth days at an average 48 kilometres per second (108,000 mph), was named by the Romans after the fleet-footed messenger of the Gods. Venus, named for the goddess of love and beauty, was tracked scrupulously by the ancient Mayans, who planned their ritualistic life accordingly. And Mars, named after the god of war, was studied by astrologers throughout the Middle Ages, in the belief that wars could be won or lost depending on the movements of the blood-red planet.

Each world gets its distinctive colour by scattering the spectrum of light received from the Sun. The colours are the signatures of the elements each contains in its atmosphere – chemical fingerprints that will someday be compared with those collected from small, yet-undetected extrasolar planets, to see if they resemble our own.

The terrestrials are also the hot planets, separated from the giant outer worlds by the ring of rocky fragments that we call the asteroid belt. All are enveloped by solar particles: Mercury and Venus are so close to the brightness of the Sun that the Hubble Space Telescope cannot turn its sensitive eyes in their direction.

For human beings, our three terrestrial neighbours are inhospitable places. It's been suggested that humans could live about 30 seconds without a spacesuit on Mars, but that, without one, they could not survive at all on Mercury. On Venus, you wouldn't stand a chance, spacesuit or not – its surface temperatures climb past 450°C (more than 800°F), and its atmospheric pressure is 90 times that of Earth. The reason is perhaps not so much that our three companion planets are so strange. As suggested in Chapter 6, it is Earth that is the peculiar planet. We have developed in such an unusual environment that we can consider colonizing only one of the other planets, Mars – and we will do that with great difficulty.

MERCURY: the Sun's Close Companion

Mercury, known to the Sumerians long before it received its name from the Romans, is the second smallest planet in the solar system (after Pluto), its diameter 40 per cent of Earth's and just 40 per cent larger than the Moon's. It has a highly elliptical orbit – at its closest approach to the Sun, it is only 46 million kilometres (29 million miles) from the Sun; at its greatest distance about 70 million kilometres (44 million miles).

Searingly close to the Sun, Mercury's temperature can reach 430°C (800°F) during the day. Then, at night, the thermometer may plummet to -170°C (-280°F). At times, with suitable eye protection, Earthbound astronomers can see Mercury as a small dark spot transiting across the Sun. It takes no special equipment to view the planet low on the horizon as an evening star in the western sky and again as a morning star in the east, but alas, the Earth's atmosphere is so thick at that angle that it is especially difficult to see in any detail. Only by patiently layering one photograph over another, and combining

We know less about Mercury than any other planet except Pluto. With its old, cratered surface, Mercury looks similar to the Moon, but it is much denser. To make this almost-complete photomosaic, data from the Mariner 10 mission of 1974 and 1975 were refined and augmented with information from radar and optical studies.

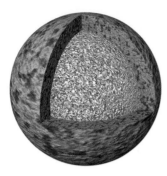

Mercury has a very high density compared to the other terrestrial planets. It is estimated that this extra density is due to a larger than usual iron core.

them with computer processing, can we gain acceptable images of Mercury from Earth. The best that we have came from the one spacecraft to have ventured near Mercury – Mariner 10.

Though Galileo was able to chart its Moon-like phases, Mercury is so small, and so near the Sun, that it has been extremely slow to give up its secrets. In the 1880s, Giovanni Schiaparelli (1835–1910) carried out a series of telescopic studies of Mercury from the Brera Observatory in Milan, and concluded that it rotated on its axis in synchrony with its orbit, so that it always presented the same side to the Sun. Percival Lowell (1855–1916), observing the planet from Flagstaff, Arizona, came to the same erroneous conclusion.

Schiaparelli also looked for features on Mercury's face, and came up with a pattern that might well be a fuzzy representation of a Chinese character, slantwise between featureless poles. In 1927, working with a more powerful telescope at the French observatory at Meudon, Eugène M. Antoniadi (1870–1844) took another crack at drawing the portrait of Mercury. It looked like a blurry representation of an Earth map, combining some of the features of eastern Asia with those of the North Atlantic. But Antoniadi's drawing did have a plenitude of spots upon it – and, by sheer coincidence, spots are what Mercury has plenty of – only they are craters. There are many, many craters on Mercury, and dark areas that look as much like Schiaparelli's calligraphy as Antoniadi's coastlines. But, while conceivably there are flat lava-rock surfaces on Mercury, the planet has absolutely none of the usual signs of volcanic activity such as cones, domes, and flow lobes.

Fifty years ago, scientists were very unsure of Mercury's mass, and estimated that its density was somewhere between those of the Moon and Mars – a very long stretch indeed, for at 5.43 grams per cubic centimetre it turned out to be denser than either.

By the end of the 1960s, backed by a huge amount of educated guesswork, space scientists had a sketchy idea of what they would find when spacecraft finally reached Mercury. Part of this was due to the large amount of information that was available about the Moon, and the predicted similarities between the two worlds. Because of their size, it made sense that these bodies didn't have enough gravity to hold on to the gases that make up an atmosphere. Without the protection of an atmosphere, meteoroids must have bombarded Mercury's surface for millions of years, and the craters should still be clearly visible because there could be no weathering effect. The apparent lack of a protective gaseous 'greenhouse' blanket also meant that the planet would be very hot in the daytime – especially so because of its closeness to the

Sun – and very cold at night. Because Mercury is so small, and yet so massive, it was assumed to have a very large core with a thin mantle and crust.

Direct observation of Mercury from Earth sites also began to yield significant results. When astronomers calculated its temperature from radio emissions they learned that its dark side is simply too warm to support Schiaparelli's and Lowell's theory. If it always faced away from the Sun, it would have to be much colder. Even if we hadn't observed the planet close enough to actually see it rotating, Mercury had to be spinning in order to warm the 'dark side.' In 1965, radar pulses from the great Arecibo dish in Puerto Rico tracked the edges of the planet with sufficient precision to see that the planet was indeed rotating – three times for every two orbits around the Sun.

Measurements of Mariner 10 photographs more or less confirmed radio observations that had shocked the science community in the early 1960s, overturning a theory that had withstood the test of time for most of a century. These indicated that Mercury rotates once every 59 Earth days – quite different from 88 days that everyone believed previously. Mariner's figure was 58.65 Earth days; the planet orbits the Sun once in every 87.97 days.

Radar measurements carried out at the California Institute of Technology (Caltech) also revealed small, radar-bright polar caps, possibly made of water and other ices, permanently preserved from melting by the shade of deep craters. Mercury is very close to the Sun, and extremely hot during the day; yet its north pole sees little of the Sun and some surmise that the insides of polar craters could be colder than -160°C (-260°F). Such temperatures are plainly cold enough to freeze water that has seeped from under the surface, or to preserve ices delivered by impacting comets. But perhaps ice is not responsible for the existence of those radar-bright polar caps. It is difficult to understand how ice could survive in those conditions, even in the permanently shaded regions. Conceivably some other substance – such as sulphur-rich compounds – could have caused the bright reflections.

The first spacecraft to use the gravitational attraction of one planet to reach another, Mariner 10 was launched on November 3, 1973, and zoomed on a photographic mission around Venus before its first rendezvous with Mercury on March 29, 1974. More than 2,700 pictures were taken, imaging 45 per cent of Mercury's surface. As suspected, these showed an intensely cratered surface reminiscent of the Moon, but with the impact sites further separated by vast plains that, because Mercury's gravity is so much stronger, are relatively uncluttered by crater debris. The craters are all too small to have been seen from Earth by Antoniadi or anyone else.

Space and the Human Body

Someday, humans will climb from a spaceship, walk across the red sands of Mars to their ready-made life-support and laboratory module, and set up house. It will be an indelibly significant event in human history. For the first time, people will reside upon one of our sister planets, where they can look for water and life, and explore the volcanic highlands, the cratered plains, and the awesome depths of the Valles Marineris.

No one knows how long we will have to wait for this event. The technology is all but ready for that first flight. But are humans ready? In our hearts and minds the answer is yes. Yet our bodies and minds are fine-tuned for life on Earth. To venture safely to another planet requires precise knowledge of the physiological effects of weightlessness during travel and reduced gravity while on Mars – plus the effects of radiation, confinement, isolation, and the dynamics of small groups.

Experience with the Apollo program, the early Soviet Salyut space stations, the U.S Skylab, Mir (which operated for 15 years, from 1986 to 2001), the Space Shuttle, and the new International Space Station gives us clues to the problems that face future Mars explorers. Weightlessness and low-gravity environments pose severe challenges to the human body. We know that additional threats exist on Mars, particularly solar storms that will shower explorers with radiation, and which will require use of warning sensors and shelters, protective clothing, and perhaps even drugs that moderate radiation effects.

From experience we know that long-term exposure to microgravity can trigger a number of detrimental physiological responses – some severe, some easily reversible. Up to 80 per cent of astronauts suffer space sickness, a sensory conflict that occurs when receptors in the inner ear no longer receive the cues that give the brain information on direction and gravity. Dizziness occurs, and hand-eye co-ordination, posture, and balance all suffer as a result. There is also a long-term deterioration of the immune system, under which long-dormant viruses can re-activate.

One of the most remarkable and dangerous results of weightlessness is bone deterioration – a loss of 1–3 per cent of bone density per month, even with regular exercise. It's a classic use-it-or-lose-it situation. The stress of normal movement in Earth's gravity stimulates bone formation and maintains bone density. Without it, bone loss and demineralization occur rapidly.

For the same reason, muscles atrophy when the body is subject to reduced gravity. But while most muscles can be exercised in space, this is not the case

Eyes become
main way to
sense motion

Loss of blood
plasma creates
temporary anaemia
on return to Earth

Kidney filtration
rate increases:
bone loss
may cause
kidney stones

Fluid redistribution
shrinks legs

Weight-bearing
bones and muscles
deteriorate

Touch and
pressure sensors
register no
downward force

Changed sensory
input confuses
brain, causing
occasional
disorientation

Otoliths in inner
ear respond
differently
to motion

Fluid redistribution
causes head
congestion and
puffy face

Higher radiation
doses may increase
cancer risk

for bone. The peak forces that we get when jogging or simply walking down stairs are impossible to obtain in space without the aid of artificial gravity.

Then there's dehydration. When freed from Earth's gravitational field, bodily fluids shift from the lower body to the head and upper body. This confuses the brain's regulatory system – more fluid in the upper body normally means there's too much fluid in the entire body. So the brain does what would normally be the right thing – it pulls the switch for the excretion of fluids. Result: a loss of about one third of body fluids.

Zero gravity is also hazardous to the cardiovascular system. While the heart normally must fight gravitational pressure to keep blood flowing, this need is removed under weightlessness, and the heart slows its pace and shrinks as a result. Another low-gravity effect is 'space anaemia,' where the production of red blood cells is reduced. No one yet knows why this occurs, but again it may be a reaction to decrease in exertion.

Other relatively minor problems experienced by astronauts include sleep deprivation – which can cause a dramatic increase in human error – weight loss, flatulence, and nasal congestion.

Not surprisingly, the main remedy or cure for these problems is frequent, strenuous exercise, a time-consuming regimen but nevertheless essential to counteract both bone and muscle deterioration.

The interplanetary astronauts' prescription may well call for four hours' exercise a day for three days, followed by a day of rest, and so on. Alternatively, Heiko Hecht of Massachusetts Institute of Technology's Man Vehicle Lab may have a better idea. 'Exercise by itself will not do the trick,' he says. 'We will have to implement artificial gravity – that is, exercise while being centrifuged. This will also reduce the need for exercise to one to two hours a day.'

Clearly the Mars experience will be exciting, but it will not all be fun. Maybe the explorers will sleep in mini-centrifuges. Just staying calm, connected, and decent will be difficult. There will be a psychological need for calls home (a message will take between three and twenty-two minutes to get to or from Earth). And there will be need for privacy. Cleaning too will be a problem, for there will be dust literally everywhere. The questions of getting food to our Mars colonists, as well as clean clothing, is another pesky, and expensive, problem.

Humans are not well designed for living in a weightless or low-gravity environment, but that won't stop us exploring Mars. Indeed, some intrepid explorers, including members of the Mars Society, are ready to venture there before many of the physiological questions are answered. As Kepler wrote in a

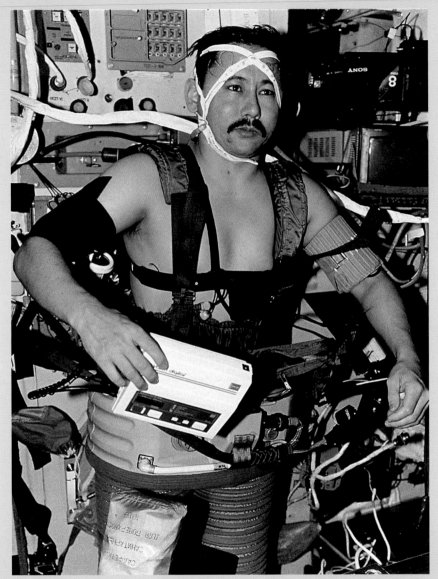

Cosmonaut Talgat Musabyev aboard the Russian space station Mir using equipment designed to prepare his body for the effects of gravity when he returns to Earth. The 'chibis' system creates a partial vacuum around the cosmonaut's legs, drawing blood down into the limbs as it would be by gravity on Earth. This equipment is used daily for two weeks prior to re-entry.

remarkable letter to Galileo in 1609, 'Ships and sails proper for the heavenly air should be fashioned. Then there will also be people, who do not shrink from the dreary vastness of space.'

The planet's gravitational pull on the spacecraft was measured to indicate its mass, which turned out to be about one-eighteenth that of the Earth. An estimated 70 per cent of the mass is represented by the gargantuan bulk of iron (with a little nickel) that underlies the thin silicaceous crust and mantle.

Two big surprises were the discoveries that Mercury has a very thin atmosphere, and an equally tenuous magnetic field, only 1 per cent as strong as the field possessed by Earth.

The Making of Mercury

Recent measurements support the idea that Mercury formed in much the same way as Earth. In its early aeons, Mercury's surface was bombarded with debris from around the nascent solar system; then a resurfacing of kinds occurred as lava upwelled from deeper regions and covered the original crust. More bombardments followed, less severe now because the supply of nebular debris was thinning. Plains formed between the craters, and the planet began to cool, causing the core to contract. As a result of this contraction, the silicate crust fractured, resulting in a folded, faulted landscape. Next, the lava flows returned and produced new flatlands – so-called smooth plains – between the higher terrain. A dusty surface or 'regolith,' much like the Moon's, churned into being as the planet was relentlessly bombarded by micrometeorites. A few large meteorites have slammed into the dust, producing large craters with bright rays stretching away from them and into the wide expanses of regolith.

Perhaps Earth looked roughly like Mercury a few billion years ago. Unlike Mercury, however, Earth continued to develop and change right up to the present day. Virtually devoid of an atmosphere and characterized by extreme heat and cold, Mercury apparently has never been flooded by water, and no plate tectonics exist by which the landscape can be modified.

Why are the outer crust and mantle so thin? A huge impact, much like the one that many believe carved out the Moon from the infant Earth, could have stripped much of Mercury's mantle. If Mercury originally formed further from the Sun, the collision could also have pushed it inward.

Suppose we were able to inhabit the body of a creature adapted to Mercury's harsh environment and explore that alien terrain. What would we find? To begin with, against that moonless black sky (a substantial atmosphere is required to paint it blue), the Sun is absolutely immense, two and a half times larger than when viewed from Earth. We might also see the yellowish disc of Venus and the blue disc of Earth.

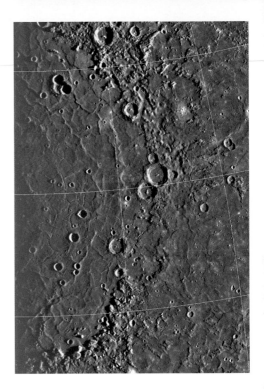

If we were fortunate, we would find ourselves near the Caloris basin, which at 1,300 kilometres (800 miles) across is the planet's largest crater. It is more than a thousand times larger than the Barringer Meteor Crater in Arizona, which is 1,200 metres (about 4,000 ft) in diameter. This great crater resulted from a truly cataclysmic event – an impact by an object that must have been about 150 kilometres (95 miles) across, rocketing in from space. Mercury was thoroughly shaken as the meteoroid penetrated its crust, producing not just one but three concentric mountain rings 3 kilometres (nearly 2 miles) high. As chunks of debris flew out of the great hole, landing 600 to 800 kilometres (375 to 500 miles) away, seismic waves thudded all the way through the planet. The result was what Mercury specialists call 'weird terrain' – a region of jumbled hills and depressions and broken crater rims, on the opposite side of the planet, similar to places on the Moon that lie opposite large impact sites. All around the Caloris basin there are smooth plains, young by comparison with the rest of the landscape, and with fewer craters, showing where lava welled up to heal the surface.

We also find smooth plains inside the crater, separated into more or less concentric patterns by long, wide ridges and broken terrain. The largest ridges are monstrous – rising to as much as 300 metres (1,000 ft) into that black sky, each would easily accommodate the area of Manhattan Island, being several hundred kilometres long and a few kilometres wide. The rim of Caloris Basin is a ring of gigantic, misshapen blocks, as much as 2 kilometres (1.25 miles) high and 50 kilometres (31 miles) square, with sheer cliffs on the inward side. Smaller concentric rings occur outside the main crater, as if ripples had spread from the original impact point and frozen in time.

An interesting place to think about, an impossible place to visit. Writing about Mercury, Venus, and Mars, Michael Seeds was moved to quote from Robert Service's *The Cremation of Sam McGee*: 'There wasn't a breath in that land of death.'

More information about Mercury's crust and atmosphere will be returned by NASA's MESSENGER spacecraft in 2009 and 2010. MESSENGER – an acronym for MErcury Surface, Space ENvironment, GEochemistry and Ranging – will make two brief reconnoitres of Mercury, then settle into orbit for one year. This mission should confirm how much volcanoes have shaped Mercury's surface, and tell us if ice exists in those shadows at the poles. It's also hoped that this

The Caloris basin, an enormous impact scar exceeding 1,300 kilometres (800 miles) in diameter, is one of Mercury's most distinctive features, and one of the largest basins in the solar sytem.

Mercury, Relativity, and Space Warps

For many decades, scientists puzzled over the Case of Mercury's Strange Orbit. Kepler's and Newton's laws showed exactly what path Mercury should travel around the Sun. It would circle in a perfect ellipse, except that the gravitational influences of other bodies would pull the ellipse around, so the most distant point from the Sun (the aphelion) would slowly circle the Sun. Its orbit would be slightly above the plane of the Earth's path around the Sun (the ecliptic) at some points, below it at others. But Mercury's orbit, just didn't square with Kepler and Newton. The rotation of its perihelion was about 0.7 per cent faster than it should be, and this disturbed astronomers greatly.

The discrepancy could have been explained if another planet were in the vicinity, exerting a gravitational force, or if, as proposed in 1855 by Urbain LeVerrier, discoverer of Neptune, there were a second asteroid belt between Mercury and the Sun. This idea gained some credibility when a French amateur astronomer said he had seen an inner planet, whereupon LeVerrier said fine, this must be one of the bigger asteroids within the belt: let's call it Vulcan.

Then, in 1915, Albert Einstein explained gravity with his general theory of relativity. He predicted that the gravity of a very massive object such as the Sun would distort the shape of space itself. This phenomenon would be enough to account for the difference between Mercury's orbit, and what could be calculated from Kepler's law. Einstein's theory also predicted that the distortion of space would deflect the path of light from more distant objects, and this effect was seen in the light from distant stars appearing close to the Sun during a solar eclipse in 1919.

The story continues. As computer models became more refined, the data again became confusing. Scientists had to put something back into the orbit between Mercury and the Sun to make their thermal models fit. These objects haven't been seen yet. But they do have a name – Vulcanoids.

spacecraft, packed with cameras and spectrometers, will uncover new information on how planets form and why some (like Mercury and Earth) have retained their magnetic fields while others have lost theirs.

Following the MESSENGER mission, ESA expects to probe Mercury's magnetic field with the spacecraft BepiColombo. Analysis of charged particles trapped by the field may indicate what kind of metal core is responsible for generating it. This ambitious project will include a main orbiter, an orbiting magnetospheric 'subsatellite,' and a lander. Partially powered by ion propulsion, it will take advantage of gravity assists at Venus and Mercury before settling into a one-year orbit. Its mission includes mapping, imaging, gravitational measurements, atmospheric analysis, and subsurface sampling.

VENUS:
The Clouded Planet

A half century ago, Venus was a mysterious place completely hidden by cloud, inscrutable and aloof. People were transported by the image of lush Venusian forests, conjured up by the Swedish chemist Svante August Arrhenius in 1918, only to have their fantasy crushed by Charles St. John and Seth Nicholson, a couple of Americans who a few years later pictured the planet as a harsh desert landscape, devoid of oxygen and lashed with great storms of carbon dioxide gas. By then, spectroscopic analysis of Venus's cloud deck had indicated that there was absolutely no water vapour in the Venusian atmosphere; measurements in 1932 showed carbon dioxide, and more carbon dioxide.

The fact is that no one had a clear idea of what lay beneath Venus's thick clouds, so imaginations could run wild. Galileo saw it as a featureless disc that went through phases like the moon, and we didn't know much more. In the 1950s it was still acceptable to imagine that the planet was endowed with great oceans of something other than ordinary H_2O – though whether of petroleum or soda-water no one could quite agree.

From 1956 onward, radio telescopes, which measure radio-frequency (RF) energy emitted by the objects on which they are focused, or which act as radar by bouncing RF signals off faraway targets, began to yield new information about Venus. The wonderful thing about these techniques is that many radio frequencies will penetrate cloud. The first radio signals received from the planet indicated that the surface was very hot – at least 315°C (600°F), an early estimate that was to prove conservative. Meanwhile, optical observations had placed the temperature at the top of the clouds at -40°C (-40°F).

Planetary Radius (km)

6048 6050 6052 6054 6056 6058 6060 6062

The first successful radar observations, which took place in 1961 at Goldstone, California, and Haystack, Massachusetts, showed how very slowly the planet was turning on its axis. Ground-based radar also began to reveal major surface features, like the Alpha, Beta, and Maxwell highlands – the only features on the planet that are not named after female personages. These early radio telescope observations could pick up signal changes (Doppler shifts) from the planet's edge, disproving the notion that Venus does not rotate, but rather presents the same face at all times to the Sun. Even more remarkably, they indicated that Venus spins in a different direction from the other planets – clockwise as seen from the north pole downward, though at a remarkably slow rate, taking 243 Earth days to rotate – longer than the planet's 225-day year. This is known as retrograde rotation, and may indicate that something – probably a major collision – tipped Venus upside down in its early history.

Earth-based observations of Venus are encumbered by a single immutable fact – when the planet is closest to us, it always presents us with the same view. As a result, while almost all of it had been imaged with radar, the resolution of our maps for most of the surface was severely limited.

Dramatic new information started to come our way as the result of visits by Russian and American spacecraft – in fact as the result of spirited competition that yielded a rich harvest of knowledge to astronomy, courtesy of Cold War political manoeuvring.

On August 8, 1962, Mariner 2 was released from its Earth parking orbit and eased on its way to the cloudy planet. It would be the first spacecraft to make a successful reconnaissance of another planet. The 202.8 kilogram (450 pound) craft was crowded with instrumentation – a magnetometer, particle detectors, cosmic ray detector, cosmic dust detector, solar plasma spectrometer detector, microwave radiometer and infrared radiometer. Two solar panels unfurled as the craft began its interplanetary cruise, along with antennae that would provide vital links to and from Earth. A month later, perhaps after a collision with some small object, attitude control was lost, than recovered in a three-minute operation by the craft's gyroscope. Real trouble set in on October 31, when the power from one solar panel decreased, making it necessary to turn off the science instruments. The panel recovered briefly, but two weeks later it quit completely. Fortunately, by that time Mariner 2 was close enough to the Sun to get all the power it needed from a single panel. The spacecraft flew by the planet at its closest distance of 34,773 kilometres (about 20,860 miles) on December 14, 1962, humming and clicking as its radios sent remarkable new information back to elated space scientists on Earth.

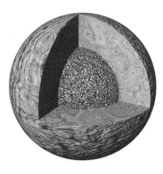

Above: Venus is nearly the same diameter and density as Earth. With geophysical measurements made from spacecraft this suggests that Venus has a lithosphere 20 to 40 kilometres (12.5 to 25 miles) thick overlying a mantle and core that may be the same size as Earth's.

Left: A global view of Venus centred on 90° longitude as revealed by a decade of radar investigation. The primary data for this image was gathered by the Magellan spacecraft which mapped 98 per cent of Venus at a resolution of 75 to 100m (246 to 328 ft). Any gaps in its coverage have been filled with data from the Arecibo radio telescope, the Russian Venera probes and the US Pioneer Venus missions.

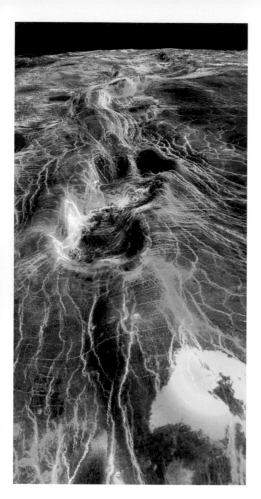

Volcanic features are abundant on Venus, including a type of volcano not seen elsewhere in the solar system. The circular depressions surrounded by fractures pictured here are called 'coronae'. They tend to occur in linear clusters along the planet's major tectonic belts and are thought to be caused by localized magma 'hot spots'. Hot, viscous magma from Venus' interior intrudes into the crust and pushes up the surface, after which cooling and contraction create the central depression and generate a pattern of concentric fractures.

Mariner 2 confirmed earlier observations that the hidden surface of Venus was hot enough to melt lead, and smothered by a high-pressure atmosphere dominated by carbon dioxide gas. Much more precise measurements were made, however, and the altitude of the cloud cover was determined to be about 60 kilometres (36 miles). This information made it possible to improve estimates of the planet's mass.

The Soviet Union then determined that Venus would be the main focus of its planetary exploration programme, and, after some failures in the early 1960s, achieved impressive results with its series of Venera spacecraft. In 1967, the Venera 4 mission included a detachable probe that entered the planet's atmosphere but quickly burned up. In 1970, Venera 7 soft-landed and became the first spacecraft to return data from the surface of another planet. In 1975, Veneras 9 and 10 became the first spacecraft to orbit Venus.

The U.S. Pioneer Venus mission radar-mapped much of the surface in December 1978–79, and released four atmospheric entry probes, two of which survived briefly before being destroyed by the terrible Venusian atmosphere. Carl Sagan recalled that one of the probes needed a sturdy window that could transmit infrared radiation, for use in a net flux radiometer. Engineers located a 13.5-carat diamond for the purpose, for which U.S. Customs levied a $12,000 import duty. Fortunately, this sum was refunded after the diamond sailed into space and Customs ruled that it was no longer available for trade on Earth.

The Soviets scored another coup in 1982, when Veneras 13 and 14 parachuted to the Venusian surface and survived the heat and pressure just long enough to send home a series of colour images and complete a soil analysis. In 1983, the Venera 15 and 16 orbiters produced a radar map of the northern quarter of Venus with a resolution of one to two kilometres, (0.6 to 1.2 miles), compared to about 25 kilometres (16 miles) for the American record.

The Americans, however, were not far behind. While the Russians were gaining praise for their accomplishments, their U.S. competitors were designing a synthetic aperture radar system that would be a true show-stopper. It was flown, with a radar altimeter, on the Magellan spacecraft, the most ambitious Venus mission to date. Gravity-mapping, which would be accomplished by determining the Doppler shift of the returned

communications signal, was a special bonus because it did not require a single gram of additional hardware.

The surface of Venus photographed by Venera 13 in March 1982.

Magellan orbited the planet between August 1990 and October 1994. Its images, made at altitudes as low as 243 kilometres (about 150 miles), picked out features as small as 120 metres (400 ft). While the Magellan images found no evidence of active tectonics, volcanoes have left their mark on 85 per cent of the surface, the rest being characterized by ranges of deformed mountains.

The spacecraft also found craters, caused by large meteorites (smaller ones would have burned up in the thick atmosphere) and scattered randomly over the surface. The strange thing about these craters was that they appeared to be very young. From what has been learned about crater formation on the Moon, it appeared that the Venerian craters had all been made within the past 500 million years – one-tenth of the planet's geological history!

How could this have happened? There are several theories. One is that, 500 million years ago, a cataclysmic collision struck Venus and forced molten material through the crust, resurfacing almost the entire planet. A more conservative idea is that the lack of plate tectonics on Venus stops the planet venting its internal heat gradually. Instead, pressure builds up beneath the crust, until eventually Venus blows its top in massive global volcanic events that resurface most of the planet.

By the end of its mission, Magellan had sent more data back to Earth than all prior planetary missions combined, mostly in the form of long thin image strips representing swaths of Venusian territory about 20 kilometres (12 miles) wide and 16,000 kilometres (10,000 miles) long. Once the cartographers had

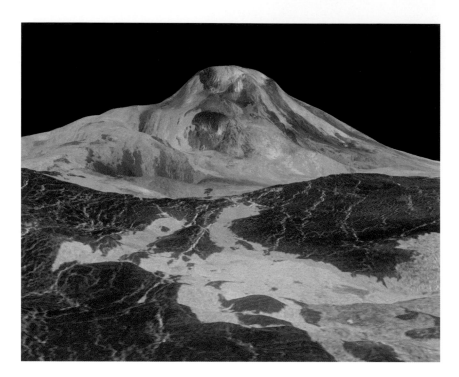

A Magellan radar image of Maat Mons, Venus' highest volcano at 8 kilometres (5 miles) high. The bright areas in the foreground are the lava flows that extend for hundreds of kilometres around the base of the volcano.

assembled these into mosaics, they had a radar image of 98 per cent of the planet, illustrating craters, plains, and circular volcanic domes that typically measure 25 kilometres (16 miles) across.

What, then, do we know of Venus, after the determined efforts of so many men and women to penetrate that cloak of cloud and unveil her secrets? A great deal more than we did before Magellan. Besides revealing the mysteries of the planet's craters, this spacecraft's images indicated that surface features can persist for hundreds of millions of years because of the lack of water and the weakness of surface winds. And the mission's gravity measurements indicated that the mantle – which surrounds an Earth-like iron core about 3,000 kilometres (1,900 miles) in diameter – is thicker than previously thought.

Sagan, normally a poetic man, called Venus 'a kind of planet-wide catastrophe,' where 'even at its high, cool clouds, Venus turns out to be a thoroughly nasty place.' A NASA report called it 'a cauldron of blistering heat and noxious gases!'

Unkind words, but one thing is certain: while our inquiries will never again be frustrated by its atmospheric cloak of secrecy, this fascinating planet remains the home of a host of intriguing mysteries.

Sarah Dunkin
Space Science Department, Rutherford Appleton Laboratory

The Starward Sisters

Each of the planets in our solar system has its own unique characteristics. All have their own enigmas, secrets that remain hidden despite our attempts to understand how each formed and evolved. The two planets closest to the Sun, Venus and Mercury, harbour many mysteries that remain unsolved, providing perplexing challenges for many scientists.

Mercury can be considered an end member of the solar system. It is the closest planet to the Sun and, apart from Pluto, the smallest. It also has the highest uncompressed density of all the planets – something completely unexpected for a planet so small. Mercury therefore plays an important part in any scientist's theory for the origin of the solar system. If a model cannot produce a Mercury, there is something seriously wrong with the model – it would be unbalanced and unstable. But how can we accurately model a planet like Mercury when we know so little about it?

The extent of our knowledge of Mercury is less than we had of the Moon before the Space Age. At that time, observing from Earth, we could see only one side of the Moon; today we have only seen 45 per cent of Mercury's surface, and this only thanks to the highly successful Mariner 10 mission in 1974–5. Thus many unknowns remain in our understanding of Mercury, and our understanding of the solar system as a whole will not be complete without a full study of this enigmatic planet.

Thirty years after Mariner 10 first flew past Mercury, NASA will launch MESSENGER, a Discovery-class mission, to get the first-ever global view of the planet. Shortly thereafter it will be followed by Europe's first-ever mission to Mercury, BepiColombo. This ESA Cornerstone mission will allow a far more comprehensive view of the planet than we have today. Crucially, it will have two orbiting spacecraft, which means the orbits and instrumentation of each can be tailored to investigate specific scientific questions. One craft will be dedicated to looking at the planet's surface, while the other will concentrate on studying its magnetosphere.

While the science to be carried out by both MESSENGER and BepiColombo will be groundbreaking in itself, we mustn't forget that the technological advances of operating a spacecraft for a year around Mercury are also significant. Thirty years ago we simply were not able to do it. Heat and radiation provide huge difficulties for the engineers designing the instrumentation for both missions. Today, radiation- and temperature-resistant materials and components are available. This means that, together, MESSENGER and BepiColombo will be able to provide answers to questions that have been taunting planetary scientists for decades.

Venus is another difficult planet to observe with Earth-based telescopes. Shrouded in a thick atmosphere, its surface cannot be seen at optical wavelengths. Therefore, detailed knowledge can best be obtained by using spacecraft in orbit around the planet.

Venus is particularly attractive to some scientists because it superficially resembles Earth. Very similar in size, mass and density, it is sometimes called Earth's twin, and in the past some thought it to have oceans of liquid water. We now know this not to be true; in reality, Venus's environment could not be more different than that of Earth.

With an atmospheric pressure 90 times that of our planet, and a surface temperature of well over 400°C (750°F), Venus is not a very nice place to be. Various radar mapping missions, the most recent being NASA's Magellan, have shown the surface to be primarily volcanic, with relatively few impact craters. Venus must have been volcanically active fairly recently, at least in geological terms, and this is an exciting prospect for scientists who are modelling the way that Venus developed.

The Venus Express mission, approved by ESA for a 2005 launch, will make it possible to study other aspects of Venus in detail. The atmosphere in particular is highly unusual, and can be said to be super-rotating, with wind speeds at the cloud tops reaching 100 metres per second (225 mph), but decreasing to only a few metres per second at the surface.

Is Venus still volcanically active? The prospect is tantalizing. We already know that Earth is not the only active body in the Solar System – Io, the moon of Jupiter, is wracked with continuing volcanic activity – but it would be an important discovery to find a planet that is still active. It is impossible to determine this from the data we now have in hand.

Although we have been exploring our solar system for decades, a multitude of mysteries remain unsolved. Mercury and Venus are just two planets that hold these mysteries, illustrating the exciting future that awaits the next generation of planetary scientists.

MARS:
The Glamour Planet

Medieval astrologers studied Mars's motions so they could advise their superiors on warfare strategies but, unfortunately for them, the planet was none too predictable. And Copernicus' 1543 theory, which proposed that planets orbit in circles around the Sun, was of course no help in predicting just where the wandering planet would move next. A real understanding of Martian motions had to wait until 1609, when Kepler discovered that Mars orbits the Sun in an ellipse, rather than a circle.

That year, Galileo confirmed with his telescope that Mars was a large sphere, not unlike Earth. In 1659, Christiaan Huygens made a drawing of Mars that included the smudge of Syrtis Major. In about 1666, Giovanni Cassini noted the planet's polar caps and was a scant three minutes off mark when he determined that it rotates once every 24 hours and 40 minutes. The German-born Sir William Herschel (1738–1822), head of a stellar scientific family, member of the Royal Society and first president of the Royal Astronomical Society, was first to discuss the thin Martian atmosphere, and its changing seasons.

When polar icecaps, clouds, and changing colour patterns were observed through the steadily improving science of telescopy, people were struck with the idea that intelligent life might reside on the Red Planet. Germany's Beer and Mädler are credited with making the first map of Mars, in 1830, but it was Giovanni Schiaparelli who in 1877 prepared a map sufficiently detailed to include names that we still use today. It included thin dark lines that he called *canali* (channels), a word that beguilingly suggested canals to English-speakers.

Uebersichts-Karte des Planeten Mars
mit seinen dunkeln linien im einfachen (nicht verdoppelten) Zustande,
beobachtet während der sechs Oppositionen von 1877-1888
von J.V.Schiaparelli.

The *canali* were also observed by Percival Lowell, while working at the 61-centimetre (24-inch) refracting telescope of the observatory he established in Flagstaff, Arizona. In his 1908 book *Mars as the Abode of Life*, Lowell theorized that the canals were made by civilized beings. Nevertheless, and despite almost universal respect for Lowell's work as an astronomer, most scientists agreed that life could not exist in a place that was so cold.

As well as noting the thin dark lines he called *canali*, Giovanni Schiaparelli also noticed that the patterns on Mars' surface appeared to change with the seasons. He wrongly assumed this to be due to seasonal changes in vegetation. We now know that Mars is swept by powerful dust storms which can alter its surface features.

Just what were the polar caps made of – if ice, what kind of ice? Herschel surmised that they were water ice, but in 1898 the Irish scientist George J. Stoney (1826–1911) said no, they were more likely frozen carbon dioxide. It was a good guess. However, in 1948 the Dutch-American astronomer Gerard Kuiper (1905–1973) argued that the caps had reflection spectra characteristic of water ice rather than of carbon dioxide. Then, in 1966, Robert Leighton and Bruce Murray of the U.S. produced evidence that atmospheric carbon dioxide would freeze at the poles. Information on atmospheric water from Mariners 6 and 7 and the Viking orbiters would show that both answers may be correct.

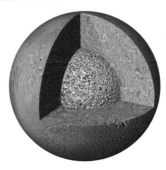

The Infant Mars

After hundreds of years of telescopic and spacecraft observation, we can surmise that, in its first billion years, Mars was torn with volcanoes, battered with meteors, and flooded and channelled by water and lava. Already, in its formative stage, it had been barraged with other great chunks of matter, perhaps so huge that they ripped away part of its young metallic core and scattered bits of the rest in the mantle and crust. Also, sometime in the immeasurable past, it is probable that a couple of small asteroids orbited a little too close, were captured by the planet's gravitational field, and became the two potato-shaped moons, Phobos and Deimos. These were first seen in 1877 by American astronomer Asaph Hall (1829–1907).

While Earth was developing a successful biosphere, Mars may have tentatively started out on the same path. At one time it probably had seas and lakes – estimates range from the equivalent of a global water layer 10 metres (33 ft) deep to that of one several kilometres deep. Eventually, though, Mars gave up the ghost, its prospects crippled by a pair of Achilles heels – a weak magnetic field and feeble gravitational pull. As a result, the solar wind blasted the planet mercilessly; lighter gases floated off into space; the atmosphere was too thin to mitigate the effects of meteor strikes. The odds are that there is no life on Mars, even though it contains plentiful water in the form of ice.

And then there's another quite different story.

Many suspect that at one time Mars had a dense carbon dioxide atmosphere, but that this thinned away as it combined with liquid water and formed carbonate rocks. The process, driven by the warmth of the greenhouse effect, continued for a few hundred million years, or more. Then the water ran out, leaving large amounts of carbonate which, if evenly spread, might form a layer 20 metres (66 ft) thick over the entire planet.

Above: Very little is known about the interior of Mars – it is usually simply modeled as a core and mantle with a thin crust, similar to Earth. Recent Pathfinder measurements have confirmed that it must have a central metallic core with a radius between 1,300 and 2,400 kilometres (800 and 1,500 miles). The absence of a magnetic field suggests that the core is probably not liquid.

Left: A view of the Red Planet synthesized from Viking Orbiter images. The north polar cap is visible at the top of the image, the great equatorial canyon system (Valles Marineris) below centre, and four huge Tharsis volcanoes to the left.

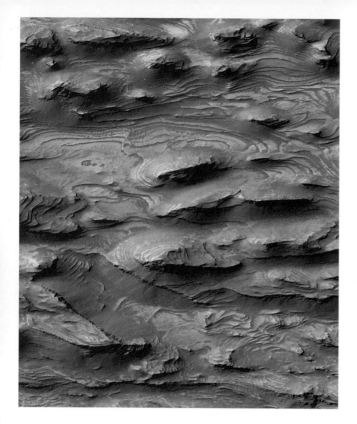

Mars Global Surveyor has revealed the presence of layered sedimentary rock on Mars. On Earth, layers patterned like this are usually only formed underwater. On Mars, these same patterns could very well indicate that the materials were deposited in a lake or shallow sea. However, it is not known for certain that these materials actually formed underwater. It is possible that there were uniquely Martian processes occurring in the distant past that would mimic the pattern of sedimentation in water.

Chris McKay of Ames Research Center's Space Science Division, and others, believe that this scenario is much more likely than the 'blowing away' of the early atmosphere by the solar wind. Earth enjoys a carbon dioxide recycling process that is driven by plate tectonics, but Mars has probably had nothing like this – it's what the scientists call a one-plate planet, with no mechanism to return CO_2 to the atmosphere. Without recycling, the lifetime of a warm early atmosphere would naturally be limited. It literally becomes petrified.

So what happened to the carbonate? It has not yet been clearly detected but, as McKay points out, carbonates are expected to be concentrated in lake basins and drainage areas. One interesting suggestion is that the layered deposits in the Valles Marineris system are made up of this material.

'The absence of spectral evidence for carbonates is troublesome,' he admits, 'but it is not a decisive argument against the present of massive carbonate deposits on Mars.' There may be reasons why carbonates are obscured from remote viewing. For example, even a thin layer of dust would hide their spectral features. Others even suggest that acid fog, composed of chlorine and sulphur gases from volcanoes, may be destroying carbonates exposed at the surface.

Mars's northern hemisphere, which consists largely of dormant or inactive volcanoes and volcanic plains, is strangely different from the southern part of the planet. There, the terrain looks more like the Moon or Mercury – except that, intriguingly, the patterns made by material thrown from the craters look more like patterns in mud than in dry regolith. Nobody knows why the two hemispheres are so different, though many suspect that part of early Mars was sheared away in a collision with some other body.

Though many photos of this small planet's surface show an uninviting landscape of sand scattered with a chaos of rocks, Mars can outdo Earth for geological drama. For starters, its highest mountain (and the highest in the solar system), Olympus Mons, peaks at 27 kilometres (17 miles) high and is 600 kilometres (375 miles) across Mount Everest is just 8,850 metres high. Its lowest point is in the Hellas Basin, 4 kilometres (2.5 miles) below the average surface.

Plans are on the drawing boards for a remotely piloted aircraft to explore the awesome Valles Marineris, which reach almost a quarter of the way around Mars, and have cliffs that tower 6 kilometres (3.75 miles) high. In scale – it measures 4,500 kilometres [2,800 miles] long and up to nearly 600 kilometres (400 miles) wide – it is comparable only with Earth's East African Rift Valley, which reaches from Jordan to Mozambique. Only Earth's submerged Mariana trench, at 11 kilometres (6.9 miles) deep, can compete with Marineris for depth.

Today, humans are considering ways of bringing life back to Mars – their own. In the near term, the planners are expected to build small, sealed human habitats supplied with oxygen extracted from Martian subsurface ice, with shelters to seek safety from solar storms, and greenhouses to grow small amounts of food. Most supplies will come from Earth. The visionaries, however, have greater things in mind. If the Martian atmosphere is too cold and dry for plants to grow outdoors, why not change the atmosphere? A nonsensical idea? Not necessarily. As with many new-technology ideas, the question will not be so much, 'Is it doable?' as 'Is it the right thing to do?' (See boxes on p.168 and p.188)

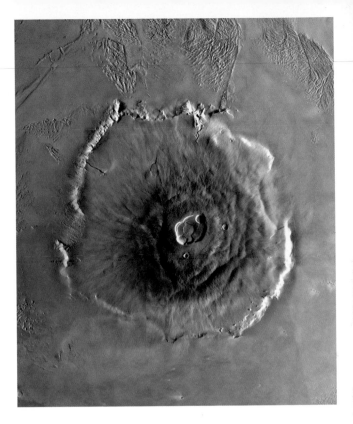

Olympus Mons, the largest volcano in the solar system, is about 600 kilometres (400 miles) in diameter and the summit caldera is 24 kilometres (15 miles) above the surrounding plain. It is as high as two Mount Everests, one on top of the other. Such large volcanoes can exist on Mars because of the low gravity and lack of surface tectonic motion.

The Red Planet at Close Range

Spacecraft finally started providing some of the answers that people on Earth were waiting for in the 1960s. Mariner 4 erased forever the 'Men on Mars' fantasy in 1965, when it flew within 10,000 kilometres (6,200 miles) of the planet's surface. Images from this pioneering spacecraft revealed a moonscape of craters, devoid of canals or other signs of civilization. And the atmosphere was indeed a tenuous affair, less than 1 per cent as thick as Earth's, and composed mainly of carbon dioxide (CO_2). Gerard Kuiper had suggested that nitrogen made up most of Mars's atmosphere, but now we know it exists only in traces, along with oxygen, carbon monoxide, and noble gases.

Terraforming

Nobody knows when humankind will be ready to send people to Mars. Yet some are already looking into ways of transforming the planet so that it would develop an Earth-like environment – 'terraforming' it so that it could support large human colonies.

Data from Mars Global Surveyor's laser altimeter have enabled us to build a virtual model of Mars (above). Using this model, we can see what Mars might look like with an atmosphere thick and warm enough to allow liquid water to exist on the surface (right). Here, the Valles Marineris has flooded to form a land-locked sea and Olympus Mons stands alone on a peninsula off the north west coast.

Chris McKay, astrophysicist and planetary scientist at NASA Ames Research Center, envisions a time when humans will produce a greenhouse effect on Mars. Chemical factories would manufacture the right combination of gases necessary to initiate a chain-reaction global warming process – it has been suggested that a mixture of five to seven fluorine compounds, mined from the regolith, could be used to sustain an Earth-like atmospheric composition and surface pressure. At -20°C (-4°F), the planet's naturally occurring carbon dioxide ice would turn to gas and mix with the human-produced gas mixture, increasing atmospheric pressure and further driving the warming process. At freezing point, the planet's stores of water ice would melt, producing liquid water and yet another greenhouse gas, water vapour. Conceivably this process could form an ocean covering as much as twenty-five per cent of the planet.

The right organism could provide the biological trigger that would begin to form a Martian biosphere, an ecosphere: if not another Eden, then a low-gravity, oxygen-lean Iowa where spacesuited people could live, farm, and enjoy the extraterrestrial life.

Enthusiasts believe the warming of Mars and the re-creation of a thick carbon dioxide atmosphere could occur in a stretch of fifty years. The sceptics say more like a thousand years. McKay suggests it would take as much as 100,000 years to convert carbon dioxide to oxygen to the point where there would be a nitrogen-oxygen atmosphere that humans could breathe.

Others say it shouldn't be done at all: it's unthinkable to consider meddling with the existence of an entire planet. But scientists are already looking into ways of making building materials from a mixture of polymer and processed Martian regolith, treated so that it is not weakened by the planet's abundance of atomic oxygen. They are building prefab Mars habitats and practising specialized geology in places that seem roughly Mars-like – such as the Dry Valleys of Antarctica, and Siberia.

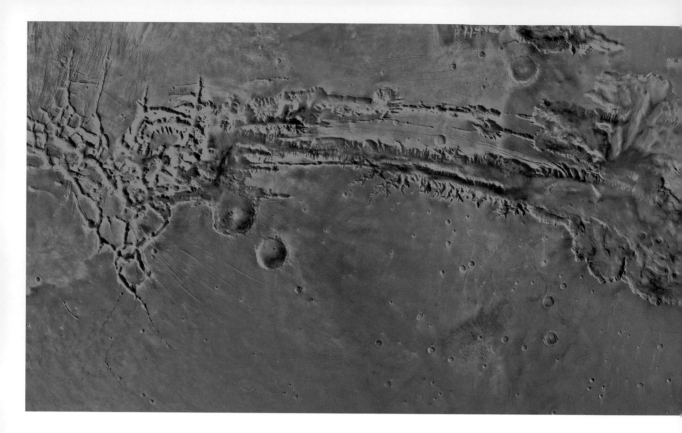

Mariners 6 and 7 were identical spacecraft that arrived in July and August, 1969, respectively, bristling with cameras and spectrometers. The cameras covered about 20 per cent of the surface, confirming that there were no canals or other signs of civilization; rather they revealed cratered highlands similar to the Moon's. Spectrometer readings showed that the thin atmosphere was almost all carbon dioxide, but the ice of the northern cap was almost certainly water because its temperature was well above the frost point of carbon dioxide. The southern cap, meanwhile, was colder and probably host to some carbon dioxide ice.

Mariner 9, the first spacecraft ever to orbit another planet, arrived at Mars in 1971. More than twice the mass of its predecessors, it was equipped with colour cameras and a fresh set of instruments for probing the planet's surface and atmosphere. It produced the first detailed global view of Mars, photographing its channels and volcanoes for the first time. Whereas earlier spacecraft saw the cratered land of the southern hemisphere, the new orbiter mapped the more dramatic features of the north and went on to image the two moons. Mariners 6 and 7 netted about 200 images, but Mariner 9 sent home 7,329, covering more than 80 per cent of the planet's surface.

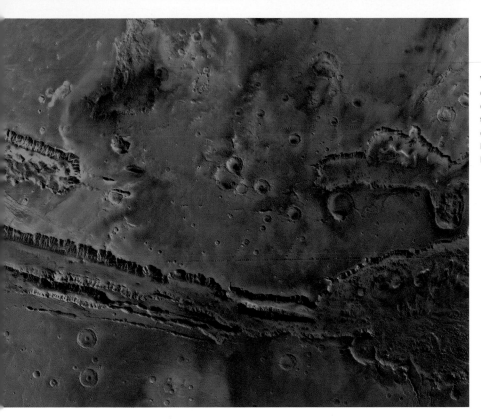

Valles Marineris, the 'Grand Canyon' of Mars, reaches almost a quarter of the way around the planet. Thought to be formed by tectonic rift in the crust when the planet cooled, Valles Marineris was subsequently enlarged by erosional forces.

Data from the Soviet Mars 5 mission, which entered orbit around the planet in 1974, suggested that the planet had a very weak magnetic field. More than two decades later, data from Mars Global Surveyor showed that, while the planet does not have a global magnetic field like Earth's, it does have localized fields strong enough to protect some parts of the surface from the solar wind.

Perhaps all of Mars, like Earth, once had a protective magnetic umbrella, but lost it gradually as the molten core cooled and solidified. As asteroid and comet impacts tore away chunks of its surface, the atmosphere would be blasted away by the solar wind and features like flowing water would disappear forever.

The twin Viking probes, each equipped with a spider-like lander, established Mars orbit in June and August, 1976, respectively. They found evidence of ground frost, detected water clouds, and imaged channels indicating that water once existed on the surface of Mars. However, the landers took photos and saw no signs of Earth-like life, even after they scooped up samples of Martian soil, soaked it, and fed it with nutrients. A few experiments seemed to indicate living organisms, but this could have been a result of the soil's mineralogical properties. For the life-on-Mars believers, it was a letdown. (More bad news came in a paper published in 2000. Free oxygen ions destroy

David Darling
Astronomer and Author, *Life Everywhere*

Life Abundant

Mars Global Surveyor's high-resolution images indicate that at some point in the past water flowed on Mars. Here, gullies on the wall of a meteor crater are thought to represent erosion by running water.

Scientific optimism about the chances for finding life on other worlds in the Solar System has climbed steadily over the past couple of decades, due to various discoveries made both here on Earth and by probes studying Mars and Europa. The astonishing diversity of terrestrial extremophiles – organisms that thrive in what appear to be the most hostile conditions – has hugely expanded the range of places where life might be expected to survive elsewhere. Evidence also continues to build that life on Earth began remarkably early, around four billion years ago, when our planet was still being heavily bombarded with space debris. This suggests that getting life started is not difficult and that, once it forms, it can swiftly adapt to many different environments both on and below the surface.

A popular belief held by astrobiologists today is that life may spring up almost anywhere given the availability of three key ingredients: a sufficient concentration of organic (carbon-based) chemicals, a biologically useful source of energy, and liquid water. All three now appear to be much commoner than previously supposed. Organic matter is contained within comets and some types of asteroid and is known to be able to survive cosmic impacts; therefore, some of the basic building blocks for life, including amino acids, are routinely delivered from space to the surface of young worlds, adding to whatever native carbonaceous material is present. Suitable energy sources for life are abundant, especially in view of the discovery on Earth of deep-dwelling microbes that live entirely independently of sunlight. The major stumbling block to biology on our neighbouring worlds had always seemed to be a lack of liquid water. But this constraint too has gradually relaxed.

Of all the possible abodes for extraterrestrial life, Mars is the traditional favourite. The Viking project's failure in 1976 to find organic matter in the Martian soil – despite ambiguous results from the landers' other life-seeking experiments – seriously dented hopes that the Red Planet had ever been biologically active. However, our picture of Mars has changed dramatically over the past few years thanks to data collected by the Mars Global Surveyor (MGS) and, more recently, by the Mars Odyssey probe.

High-resolution images from MGS, which entered Martian orbit in 1997, support the view that Mars once had extensive surface water, including lakes and, probably, a shallow ocean that covered much of the northern hemisphere.

It is clear that Mars was both reasonably warm and wet between 4 and 3.5 billion years ago – sufficient time for primitive organisms to have developed. But does life exist there today? Intriguing pictures sent back by MGS show stream-like features cascading down the sides of more than a hundred craters and canyon walls. All of these features consist of a deep, narrow channel with a collapsed region on the crater or valley rim, and a fan-like area of debris at the lower end. They appear to have been made recently – within the past million years – by some liquid erupting in a sudden flood just below the surface, high on a steep slope, undermining and collapsing the ground there, and then racing downhill carrying material with it. Among the theories to explain these 'gully-washers' is that liquid water still exists at shallows depths on Mars. Whether or not this is the case, water may very well survive in deep aquifers. Such structures are among the likeliest places to search for extant Martian microbes.

Several pieces of Mars have fortuitously found their way to Earth, having been blasted into space by colliding asteroids and then later swept up by our planet. Controversy still surrounds what some scientists claim are biogenic traces in these meteorites. The fact that such interplanetary exchange takes place at all, however, makes it feasible that primitive life – for example, in the form of dormant bacterial spores – might travel from one world to another. It is even possible that our own remote ancestors came from Mars!

Astrobiological interest also extends to the outer solar system. Observations of Jupiter's second Galilean moon, Europa, strongly suggest that a watery ocean lies beneath its icy surface. Saturn's giant moon, Titan, rich in organics yet too cold to hold to harbour life, prompts us to speculate how far along the road to biology an environment like this can progress.

Some answers to these outstanding issues may come soon. The Huygens probe, carried by Cassini, is scheduled to parachute onto Titan's surface in 2005. Meanwhile, in late 2003, the European Mars Express will carry the tiny British-built Beagle lander to Mars to sniff the atmosphere and sample the soil for traces of Martian life.

A panoramic view from the Mars Pathfinder lander.

organic molecules – and they are produced in abundance by intense ultraviolet radiation in the thin Martian atmosphere. By mixing with regolith minerals, they can produce superoxides that are deadly to life. If there is life on Mars, then, it may be limited to areas submerged below the superoxide-rich layer.)

Discoveries from Mariners 6 and 7 and Viking accelerated the search for evidence of water on Mars. Though free-standing water is a thing of the distant past, it was clear now that its remnants survive in frost, unmeasured sub-surface reserves, and a few clouds.

Liquid water can't exist on Mars because the atmosphere is so thin – with an atmospheric density just one per cent of Earth's, a visiting astronaut's glass of Perrier would instantly vaporize. And Mars doesn't offer much in the way of water vapour, either. It's estimated that if it were all brought together and solidified, it could form an iceberg-sized mass with a volume of about 1.3 cubic kilometres (0.3 cubic miles).

Underground water almost certainly exists, probably at 100 to 400 metres (330 to 1,300 feet) below the surface. Those layers could contain carbonates, providing evidence that liquid water did indeed exist on the planet's surface. They might even contain fossils left by ancient Martian bacteria.

From the wind-down of Viking in 1982 until 1996, no successful spacecraft missions reached Mars. Finally, in July 1997, Mars Pathfinder touched down. This mission, which deployed a robotic microrover, Sojourner, demonstrated a low-cost method of delivering science instruments to the surface.

A northern hemisphere crater in the Martian spring, showing frost that has built up over the winter.

Then, beginning in September 1997, Mars Global Surveyor (MGS) began to amaze us with astounding new discoveries. It watched the weather, with its whirlwinds and huge dust storms that suddenly appear, then just as quickly disappear. It mapped the Valles Marineris, and showed that the planet's upper crust is layered. Frost patterns indicated seasonal change. Water had apparently emerged at the surface – perhaps recently. It became steadily more difficult to argue that water (or some other liquid or mix of liquids) had not carved channels into the crust. 'Gully washers' found in MGS imagery appear to be geologically young, at a million years old or less. But nobody knows what would have caused these rivers and rivulets to flow.

In 1999, James W. Head III of Brown University announced that MGS data suggested 'the ancient shoreline of a large standing body of water present in middle Mars history.' A year later, MGS produced remarkable images that looked like layers of rock laid down by receding lakes and seas. Imaging specialist Michael Malin, until then doubtful that water had existed on Mars, called it 'irrefutable evidence that sedimentary rocks are widespread on Mars.'

Controversy thrived. Nick Hoffman of Australia's LaTrobe University produced a model in which liquid carbon dioxide, squeezed to the surface by sinking regolith, created water-like erosion, with 'no need for a warm and wet epoch nor an ocean.'

But the MGS camera system provided one mindbending image after another. They seemed familiar – rocks, sand dunes, gullies and volcanoes – and yet they were not. The geological features of Mars took form under different circumstances from those of Earth, in a very different environment and over different timescales. Many of them are unexplained, and are likely to remain so until human geologists begin work on the Red Planet.

Global Surveyor finished its primary mapping mission in early 2001, having stirred a beehive of excitement and speculation that lakes and seas, and perhaps even marine life, once existed on Mars. That year, a study of data from the Far Ultraviolet Spectroscopic Explorer (FUSE) satellite showed that Mars once had more water, proportionally, than Earth.

Interest in water on Mars redoubled when Mars Odyssey arrived in February 2002. This spacecraft almost immediately detected significant amounts of hydrogen, indicative of water, in the planet's south polar region. Data from its gamma ray spectrometer indicated that the region contains enough water to fill Lake Michigan twice over, trapped in a layer of muddy sub-surface ice in the top metre (3 ft) of regolith. University of Arizona scientists jubilantly announced that the ice appears to be at least fifty per cent water. Meanwhile, images of the Cerberus Plains north of the equator indicated that liquid water may have surged to the surface from the planet's interior as recently as 10 million years ago.

NASA commentators joked that the dirty-ice discovery might be 'the tip of the iceberg.' And they may be right.

More robot explorers would soon be en route – new NASA orbiters and landers; the Japanese orbiter Nozomi; and ESA's Mars Express with its British-built Beagle lander. The first mission to return regolith samples will probably occur sometime after 2010.

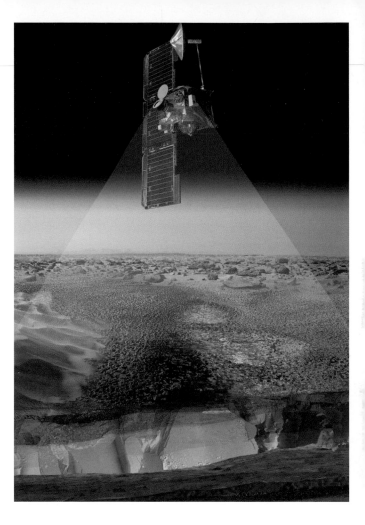

Using its gamma ray spectrometer, NASA's Mars Odyssey spacecraft detected enough water ice lying within a metre (three feet) of the Martian surface to fill Lake Michigan twice over. It is not known whether the ice-rich zone continues below this depth. This view is not to scale, as Mars Odyssey's observations are made from an altitude of 400 kilometres (250 miles).

Considering the innate inhospitality of Mercury and Venus, Mars is naturally the most attractive of the terrestrial planets for human exploration and future colonization. The moons of Jupiter, despite their apparent wealth of water, are not likely to give much competition in this regard. Arthur C. Clarke says that Mars is where the action will be in the twenty-first century. 'It is the only world in the solar system on which there is a strong probability of finding Life Past, and perhaps even Life Present,' he wrote. 'Also, we can reach it – and survive on it – with technologies which are available today, or which we can acquire in the very near future.' Some believe we already have evidence of archaic Martian life, in the form of Martian meteorite ALH84001 (see p. 224).

And then again, perhaps the greatest discoveries of the solar system await us in the huge systems of Jupiter and Saturn. As Carl Sagan put it, 'Somewhere, something incredible is waiting to be known.'

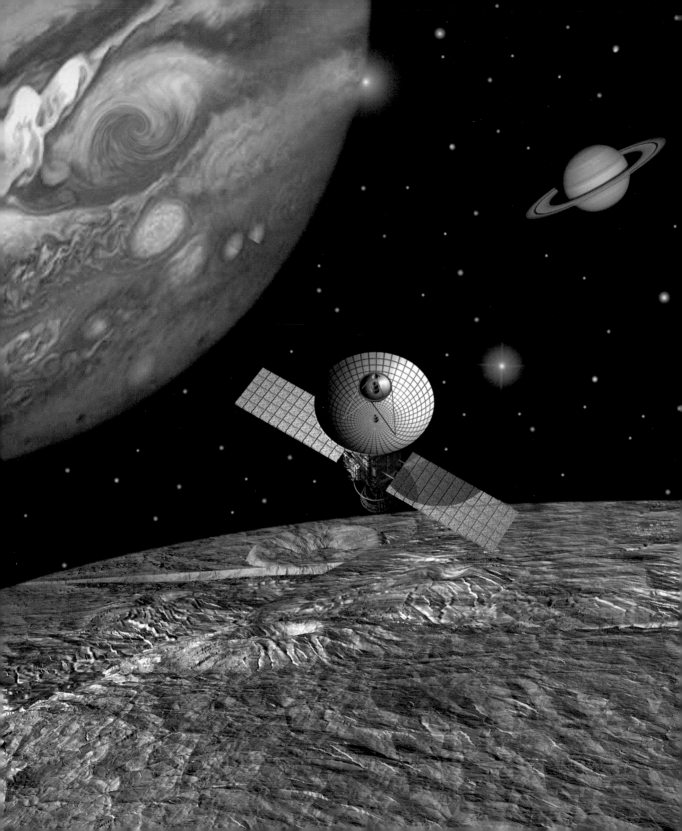

Twin Giants

'The waters gleamed, the sky burned with gold, but all was rich and dim, and his eyes fed upon it undazzled and unaching. The very names of green and gold, which he used perforce in describing the scene, are too harsh for the tenderness, the muted iridescence of that warm, maternal, delicately gorgeous world.'

C.S. Lewis

The last two decades of the twentieth century were a fireworks show of planetary discovery. Mars was a show-stopper, but perhaps the most dramatic of these findings came from Jupiter and Saturn and their moons. Nothing in the known solar system, except for the Sun itself, comes near to matching the drama of these planetary giants. And many great mysteries remain. Certain moons have revealed their surfaces to planetary spacecraft. But as for the planets themselves – no one is sure of their composition or even whether they have solid surfaces beneath their thick mantles of cloud.

In the early days of the solar system – the first 100 million years or so – two planets grew so fast and massive that they were able to attract huge amounts of gas from the solar nebula. Jupiter and Saturn soon established themselves as giants of the system, planetary musclemen that will determine, for example, whether and when another ecology-crippling asteroid will strike Earth.

These planets contain more than 90 per cent hydrogen, and most of the rest is helium – along with small amounts of other gases, such as methane. They owe their composition not only to their original makeup but also to the asteroids and comets and absorb with their enormous gravity.

It's now believed that the orbits of the giant planets changed gradually after their initial formation, as minor planets were either swept out of the early solar system or captured by the gravitational fields of the larger planets. This theory, which explains the development of a dynamic balance within the solar system, holds that Jupiter's orbit shrunk slightly as the orbits of Saturn, Uranus, and Neptune enlarged.

Jupiter and Saturn gained their dominance during the early nebular 'cleanup phase,' when the planets were just forming and debris was tumbling together into the system we know today. The process continues: another step occurs every time one of the giants grabs a chunk of comet or asteroid, and either holds on to it or tosses it into a more remote solar orbit.

JUPITER:
The Giant

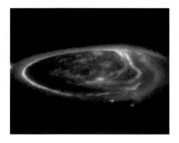

When the two Voyager spacecraft raced by Jupiter in 1979 they found a brightly coloured, hurricane-wracked atmosphere punctuated by the famous Great Red Spot, boiling with energy that was set afire when the solar system began. Aurorae, including ultraviolet displays that were undetected by the earlier Pioneer missions, graced the poles. Rings (probably composed of dust blasted from the surfaces of tiny moons by meteorites) were revealed by three separate planetary images as Voyager 1 passed through Jupiter's equatorial plane – a delightful surprise, for until then only Saturn and Uranus had been known to have rings. The 39 known Jovian moons, of which only eight had been viewed by Voyager, include several that are wonderlands in themselves.

Above: An ultraviolet image of Jupiter's aurora taken by the Hubble Space Telescope. Although the aurora resembles the same phenomenon that crowns Earth's polar regions, this image shows unique emissions from the magnetic tracks or 'footprints' of three of Jupiter's largest moons, Io (along the left hand limb), Ganymede (near the center), and Europa (just below and to the right of Ganymede's auroral footprint).

Right: A Cassini eye view of Jupiter. The shadow is cast by Jupiter's moon Europa.

'The planet's face is an immense canvas splashed with brick red, straw yellow, snow white and gun-metal blue,' announced the *Christian Science Monitor*. 'Features swirl and collide, form and break up, brighten and fade in a stately and intricate choreography, its fastest tempo measured in hours and its longest in centuries.'

Jupiter, so large that some compare it with a failed sun, always appears brighter from Earth than even the brightest star, Sirius. Because it is so large – and relatively close – it has been studied ever since the first crude telescopes came into being. Galileo discovered the four largest of its satellites in 1610; in 1630 Nicolas Zucchi and Daniel Bartoli observed the lateral belts (dark bands) and zones (bright bands); the Great Red Spot was described in 1664 by both Italy's Giovanni Cassini (1625–1712) and England's Robert Hooke (1635–1703).

Jupiter is 318 times more massive than Earth, an irresistible attraction for astronomers, cosmologists, and indeed everyone whose gaze lifts from earthly things to look out into the universe. NASA scientists were quite naturally no exception in their enthusiasm to explore this magnificent sunlet.

Any scientist will tell you that, when it comes to gathering data, it's impossible to improve on up-close inspection. Still, it's remarkable how many Jovian mysteries were revealed by scientists working with various kinds of

An anti-cyclonic storm akin to an Earth hurricane, Jupiter's Red Spot has raged throughout the four centuries that humans have been observing it with telescopes. The distance from top to bottom of this Voyager picture is 24,000 kilometres (15,000 miles). The white area below the Red Spot is one of three white ovals that were observed to form in 1938; they move around Jupiter at a different velocity from the Red Spot.

telescopes at home on Earth, decades before the Hubble telescope went into action. The Voyagers detected huge lightning storms on Jupiter, but this was no surprise. Back in 1955 scientists discovered that storms, recognisable only by the intensity of radio noise they caused, were occurring on Jupiter. They truly seemed like Jove's mythological thunderbolts, causing low growls, cracks and roars when played through a sound system.

Some of the cloud-top lightning traces detected by Voyager 2 were considered to be 'superbolts,' between one hundred and one thousand times more intense then a typical lightning flash in Earth's atmosphere. And the lightning storms never abated – up to 80,000 lightning flashes occur every second on Jupiter, compared to a mere 100 per second on Earth.

Scientists believe that temperatures rise from -158°C (-252°F) at the cloud tops to more than 20,000°C (36,000°F) at the planet's centre. The source of all this heat is believed to be partly residual heat left over from planet's development (the same energy that keeps the interior of our planet molten), and partly a kind of frictional heat engine that works by the gravitational separation of hydrogen from helium. Remarkably, Jupiter and the other gas giants may retain all of their original shares of the solar nebula that gave birth to the planetary system!

Dissecting Jupiter

The 4,000-kilometre-deep (2,500-mile) Jovian atmosphere has a very complex structure. This was clear even from early studies that were made without the benefit of spacecraft instrumentation. These indicated that the planet's visible cloud layer is composed of clouds of ammonia and ammonia crystals, suspended in an atmosphere made up of hydrogen and helium mixed with something else that gives it its colour. The 'something else' was once presumed to be methane and ammonia; more recent studies indicate that it is more likely to be ammonium hydrosulphide.

While we know that the colourful bands that appear in images of Jupiter are associated with different atmospheric levels – red at the highest altitudes, white a little lower, then brown, then blue – and while we know they correspond to alternating regions of east/west winds of speeds up to 480 kilometres an hour (300 mph), we still don't quite understand what whips up those winds or what causes their coloration.

The famous Red Spot retained much of its mystery even after close-up inspection by the Pioneers. What exactly is it? For nearly a century many scientists believed it was a raft of some kind, floating in the planet's atmosphere. Some reckoned that it might be an atmospheric disturbance like a cyclone. Others believed it to be organic material sparked into being by a lightning discharge – perhaps marking the first step in the development of life. As long as their observations were limited to what they could detect with earthbound instruments, nobody knew for sure, and the Red Spot was Jupiter's most tantalizing brain teaser. Eventually Voyager 1 showed that the anticyclone theory was correct – the spot is a great whorl of gases rotating at up to 100 metres per second (225 mph) as the entire structure moves slowly westward. It is about 25,000 kilometres (15,500 miles) long and half as wide. The red colour may be contributed by phosphorus and/or sulphur compounds, but we lack hard and fast evidence to confirm this. Maybe the colour does indeed come from organic compounds.

The knots of light, which have been circled in yellow in this Galileo image, represent lightning in Jupiter's atmosphere. The largest of the circled spots is over 500 kilometres (310 miles) across.

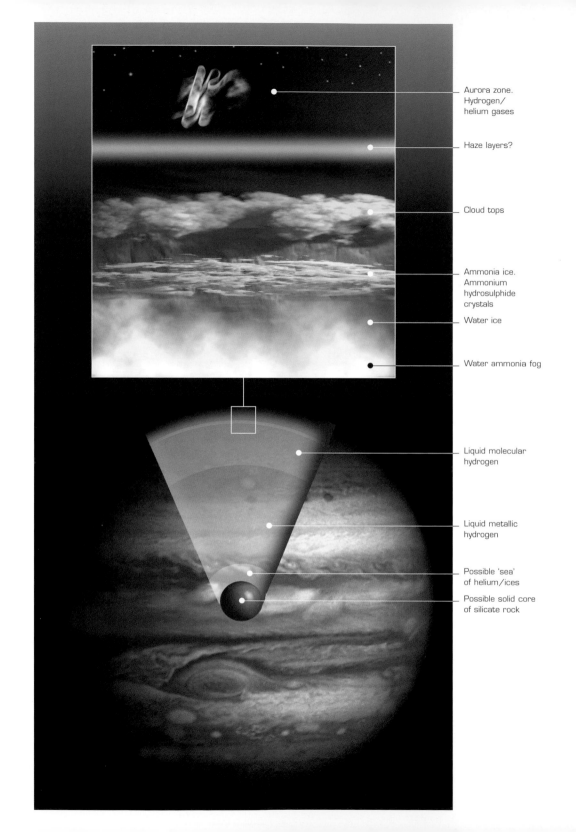

Aurora zone. Hydrogen/ helium gases

Haze layers?

Cloud tops

Ammonia ice. Ammonium hydrosulphide crystals

Water ice

Water ammonia fog

Liquid molecular hydrogen

Liquid metallic hydrogen

Possible 'sea' of helium/ices

Possible solid core of silicate rock

We were foxed by the other baffling shapes that were imaged by the Voyager cameras – for example the so-called White Ovals and White Plumes. By the late 1980s we knew that they were storms too, thunderheads of freezing ammonia gas that stand out starkly against the planet's brown haze. In 1998, astronomers watched in awe as two egg-shaped Jovian thunderheads crashed together and merged into a superstorm the diameter of Earth.

In July 1994, a 'string of pearls' of cometary fragments called Shoemaker-Levy 9 slammed into Jupiter, disrupted the planet's magnetosphere and caused huge aurorae at the poles. Just seventeen months later, on December 12, 1995, the Galileo spacecraft sent its own missile hurtling into Jovian atmosphere at 170,000 kilometres per hour (106,000 mph). The probe's discoveries were so surprising that nobody quite believed them. As expected, there were tremendous winds – so-called auroral electrojets whirl around the poles at an average 2.8 kilometres per second (6,300 miles per hour) – and great extremes of heat and cold. But Jupiter was drier than expected. There was half the expected helium. And where were the tiers of cloud? This left a lot of scientists scratching their heads. Finally, it was agreed that the probe happened to enter Jupiter at a place that in some ways was not typical of the rest of the planet.

What of the world under the clouds? After the Voyager flights, the revised model suggested that molecular hydrogen and helium exist in the outer layer, metallic liquid hydrogen in the middle layer, and a relatively small core of molten rock and ice at the centre.

Magnetospheric Marvels

Like Earth, Jupiter is girded with magnetic fields. Similar magnetic structures – magnetospheres – were discovered at Saturn by Pioneer 11 in 1979, at Uranus by Voyager 2 in 1986, and at Neptune in 1989, also by Voyager 2.

Like common bar magnets, planetary magnetic fields generally have north and south poles linked by magnetic field lines. At times, electrons and charged particles that are captured by these fields are carried to the poles, where they enter the atmosphere and trigger the brilliant displays we know as aurorae.

Because planets orbit the Sun through the unceasing outward flow of charged particles in the solar wind, planetary magnetic fields are not evenly shaped. They are like great windsocks, pressed relatively close to the planet on the sunward side by the solar wind and trailing off on the other side, driving a long spike of magnetic flux into interplanetary space.

Jupiter's magnetosphere is 10 times the diameter of the Sun, with a 'magnetotail' that probably reaches at least 5 A.U. (750 million kilometres or 470 million miles) into space – as far as Jupiter's distance from the Sun! Electric currents from Io and other Jovian moons disturb the magnetosphere, making tracks that appear like luminous contrails in ultraviolet photos from the Hubble Space Telescope's imaging spectrograph.

Magnetic fields are measured from spacecraft with instruments that resemble ordinary magnetic compasses. A magnetometer typically consists of three of these compasses, mounted at right angles to each other. The field strength is measured in each direction, then the results are combined to give the precise direction and strength of the field at the point where the measurement is made. The Voyager spacecraft had two low-field magnetometers, mounted 6 metres apart on a 13-metre boom, and two high-field magnetometers at the base of the boom.

Jovian Moons

Io is the most volcanic body in the solar system. This Galileo image captures two volcanic plumes. The plume on the edge of the moon reaches out 140 kilometres (86 miles) into space. The second plume, near the terminator (the boundary between day and night) in the centre of the image has been seen in every Galileo image with the appropriate geometry, as well as every such Voyager image acquired in 1979. It is therefore possible that the volcano beneath has been continuously active for more than 18 years.

Confronted with the great treasure-chest of images and data that resulted from our observations of the planet Jupiter, it's hard to believe that its system of moons could be even more interesting than their mother planet. Yet this may be the case. We had an inkling in 1979, when the two Voyager spacecraft sailed by all four of the Galilean satellites – Io, Europa, Ganymede, and Callisto – before proceeding their separate ways. They sent us magnificent images of these four moons as well as their tiny sister Amalthea, long before Hubble and Galileo came online. These early observations proved to be just appetizers on the menu of the magnificent mind-feast that is the Jovian satellite system.

Astronomers have often pondered the idea that Jupiter might be a failed sun – so huge, yet untouched by the cosmic ignition spark that would have turned it into a star. It even has its own solar system of huge moons, three of which may have saltwater oceans – huge bodies of water buried below tens of kilometres of ice. In early 2002 the number of Jovian moons jumped to thirty-nine, following the discovery of eleven captive asteroids, snared in the planet's gravitational field, by astronomers from the University of Hawaii and Cambridge University.

Io is the innermost of the four major moons, twenty per cent more massive than Earth's Moon, though only slightly larger. Eight extraordinary volcanoes were spotted here by Voyager 1, and we now know Io is the most active volcanic body in the solar system. Apparently Io's crust is flexed by a kind of tug-of-war

between the gravitational fields of its neighbouring moons and Jupiter itself. Intense heat is generated by friction, forming an environment of volcanoes and earthquakes that would be cataclysmic by Earth standards.

It is an infernal environment, yellow-coloured with sulphurous lakes and rivers. Sulphur-containing material from Io's largest volcano, named Pele for the Hawaiian volcano goddess, is ejected to more than 30 times the height of Everest, falling on an area the size of France. Reporting in *Science* magazine after four years of Galileo spacecraft observation, A. S. McEwen of the Lunar and Planetary Laboratory, University of Arizona, commented that 'we observed an active lava lake, an active curtain of lava, active lava flows, calderas, mountains, plateaus, and plains.' Io's mountains, like the gargantuan Tohil Mons (about 6 kilometres or 4 miles high) may result from tectonic forces, perhaps like those of the early Earth.

Radio signals, modulated by changes in the Galileo orbiter's speed as Io's gravity pulled at it during the initial fly-by, indicated that this moon has an iron core, probably surrounded by molten rock and a thin crust; a magnetometer picked up evidence that Io also has its own magnetosphere,

while imaging instrumentation confirmed that the moon is replete with red, green, and blue aurorae.

Jupiter's Galilean moons compared in size to Earth. From left to right: Europa, Io, Callisto, Ganymede and Earth.

For physical drama, Io outdoes the best special effects that the movie world could ever provide. But Europa, second of the giant moons, may, just may, offer us something more – extraterrestrial life. Voyager and Galileo spacecraft measurements indicated that Europa almost certainly has liquid water below its global icecap (Earth's Lake Vostok, in Antarctica, is similarly sealed beneath a 4-kilometre [2.5-mile] crust of ice). 'If you're looking for environments where life might take form and survive,' says Galileo project scientist Torrence Johnson, 'Europa might be a good candidate.'

This satellite, two-thirds the size of our own Moon, was depicted by Voyager 2 as smooth and scratched-looking. It also appeared whitish, showing for the first time a mantle of ice that could be 10–20 kilometres (6–12 miles) thick. What caused the scratch marks? Nobody knew for sure until the Hubble telescope unveiled a few more secrets (see p.210). And what caused the slight coloration of the ice surface? Probably sulphur, and sulphur oxides, which may have migrated from Io or upwelled from volcanoes in Europa's interior.

Intriguing Europa

By the end of last century we could visualize a Europa whose surface is cracked and crazed by shrinking ice, as underlying water is warmed by the energy of Jovian tides. Salty water sloshing back and forth produces electric currents, accounting for magnetic field variations that were detected by the Galileo spacecraft. Maybe the subsurface ocean has hydrothermal vents like those that are hidden away from sunlight deep in Earth's oceans, supporting life without photosynthesis. If the sulphur detected on Europa's surface originates from local volcanoes, this notion will begin to gain credence.

In 1995, Johns Hopkins University scientists announced that they had detected a tiny trace of molecular oxygen in Europa's atmosphere, visible in ultraviolet images from the HST. The data indicated only enough oxygen to fill a dozen large covered stadiums at Earth-like pressure, but it was oxygen nevertheless. Ions from Jupiter's magnetosphere rain down on Europa's frozen seas continuously, releasing hydrogen, oxygen, and oxygen compounds from the ice. The hydrogen floats off into space because it is too light to be captured by Europa's gravity. Oxygen remains as a very thin atmosphere, but there's not much around with which nature could perform the process of photochemistry, and the available energy from the Sun and other sources is only a tiny fraction of that on Earth. The prospect for life is dismal indeed – or is it?.

It's difficult to settle on any particular theory that might support an argument for life in Europa's ocean. The ice is believed to be some kilometres thick. It cracks and folds in a kind of icy plate tectonics, and is sometimes fractured by asteroids – one great crevasse is the length of California's famed San Andreas Fault. But none of these phenomena is believed to provide an opportunity for sunlight to reach inward to the icebound sea.

At least one group is working on a robotic mole that would burrow through Europa's ice in the search for life. Mohamed El-Genk and colleagues at the University of New Mexico believe that an instrumented package could melt its way through the ice, using long-lived energy from a radioisotopic heat source.

Do the forces of plate tectonics wrestle the crust beneath? We just don't know. Europa may have a metal core and silicate mantle similar to that of Earth or Mars. If like Earth it generates volcanic energy, perhaps there is hydrothermal activity between the mantle and ocean.

Physicist Freeman Dyson suggested that, if there is life in those oceans, hydrothermal explosions might well splash samples of it onto the ice or into orbit around Jupiter. 'We might find many unexpected surprises in the Jupiter

Top: Fractures in Europa's icy plains have brought non-ice material to the surface. This may mean that Europa harbours liquid water beneath its crust.

Bottom: The pattern of cracks and ridges on Europa's surface is remarkably similar to that seen on the ice covering the Earth's Arctic Ocean.

Ice crust

Liquid water (?)

Rocky interior

Core

ring, even if we do not find freeze-dried fish,' he said, with tongue only partly in cheek. 'We might find freeze-dried seaweed or a freeze-dried sea monster.'

Not long afterwards, astrogeophysicist Brad Dalton of Ames Research Laboratory found that spectra from the reddish-brown 'rust' around Europa's dark cracks are intriguingly similar to the spectra of heat-loving bacteria that live in the hot steam vents of Yellowstone Park.

Conceivably, extraterrestrial life could emerge from some quite exotic chemical process. Jeff Kargel and colleagues at the U.S. Geological Survey see a number of chemical pathways that could support 'viable ecosystems' similar to some of Earth's more exotic life-forms. One such is methanogenesis, a process in which hydrothermal alteration of crust minerals and carbonates produces hydrogen and carbon dioxide. Biology can then make use of this energy, combining carbon dioxide and hydrogen to produce methane and water.

Scientists have also considered whether reactions involving hydrogen and sulphur and/or iron could provide the spark of life for Europa. But the fact remains that we know too little about the 'Europan biogeochemical cycle,' if it exists, to put together a convincing theory. 'One must be careful when doing comparative planetology,' says Eric Gaidos of the University of Hawaii. 'It is not a safe assumption to use Earth as an analogy. A liquid-water ocean on Europe does not necessarily mean there is life there.'

True. But for all the doubts and caveats, there is a steadily growing hunch that life, if it can exist, will exist.

Left: Slightly smaller than Earth's Moon, Europa is the solar system's 'smoothest' object, with no geological features exceeding 1 kilometre (0.6 miles) in height. The bright feature containing a central dark spot in the lower third of the image is a young impact crater some 50 kilometres (31 miles) in diameter.

Right: A possible model of Europa's interior.

Io casts a dark shadow onto Jupiter in this Cassini image.

Ganymede's large shadow always seemed to show up on early Jupiter photographs – not surprising, since it is the largest moon in the solar system, bigger than Mercury or Pluto and three-quarters the size of Mars. The Voyagers found regions of angular shapes, streaks and gouges, with ridges and troughs that are up to 16 kilometres (10 miles) wide and several hundred kilometres long. Sharp images from the Galileo spacecraft revealed a surface crowded with young craters, punched into icy terrain that has grooves, ridges, and older craters. Impact-generated dust coats the older terrain, and there is a very large dark circular feature in the northern hemisphere. Magnetic readings indicate that Ganymede has an iron-rich core and auroral displays, like Earth. Galileo's magnetometer data also suggest that, like its sister moon Europa, it may be endowed with a subsurface saltwater ocean.

A large part of Ganymede's surface is covered by relatively young ice. How it formed is a puzzle, but that underwater ocean may be the key. Tectonic events may have been responsible.

In 1995, readings from Hubble's Faint Object Spectrograph indicated that ozone – a molecule made up of three atoms of oxygen – is being sandblasted from ice particles on Ganymede's surface. As on Europa, the 'sand' is composed of charged particles captured at high speed in Jupiter's magnetic field. This indicates that Ganymede may have a very thin oxygen atmosphere – again like Europa. In the following year, Hubble measurements showed us that Ganymede has a magnetic field, complete with auroral displays.

Younger than Ganymede and composed of lighter-weight materials, Callisto is a much-cratered, battered-looking sphere fifty per cent larger than our own Moon, pocked with craters that, oddly, form concentric rings on its left half. How did these rings form? Maybe – just maybe – they developed extremely slowly after a huge asteroid strike. The moon's spherical shape might have re-formed over many millions of years, like an incredibly thick glob of molasses after it has been pierced by an airgun pellet. During these thousands of centuries, the ripples were peppered with smaller impact craters, forming the terrain that we see today.

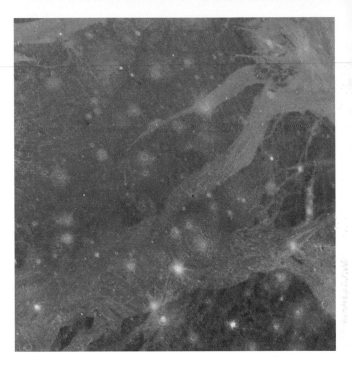

Galileo magnetometer measurements of Callisto resembled those from Europa, prompting scientists to conclude that this icy moon too, and perhaps still others in the solar system, contains a subsurface reservoir of water. Whereas water layers may be sustained by the heat of radioactivity in Callisto, tidal heating is thought likely to be more important in the case of Europa

As well as viewing the large Galilean moons, Voyager took a look at Amalthea, a roughly bean-shaped, asteroid-like moon closer to the planet than Io. It measures about 270 by 150 kilometres (170 by 95 miles), and was discovered in 1892 by Edward Barnard (1857–1923) of Yerkes Observatory. It gives off more heat than it could absorb from solar radiation, so there's some dynamic force at work here – maybe from the pounding of particles trapped in Jupiter's magnetic field, or from electrical energy generated as Amalthea travels through the field.

In the late 1980s, Galileo images revealed dust coming off Amalthea and its sister moon Thebe as they are struck by microscopic meteoroids. This material joins the so-called gossamer rings of Jupiter, and the same kind of thing apparently happens on the Jovian moons Adrastea and Metis, which shepherd another part of the rings between them. Something similar may occur around the other gas giants, and – at a very much smaller scale – might represent a model of what happened when the solar system was being born.

Ganymede is the solar system's largest moon, and is even larger than Mercury. Its surface is equally divided between relatively dark terrain, which is heavily cratered and considered to be ancient, and bright terrain, which is younger and characterized by grooves and ridges. This image clearly shows older dark terrain cut by the glacier-like flows of younger light terrain.

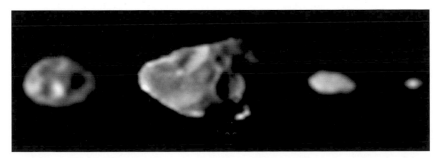

Inside the orbit of Io are Jupiter's four inner moons. From left to right are Thebe, Amalthea, Metis and Adrastea. The largest of these moons, Amalthea, is 189 kilometres (117 miles) in diameter, whereas the smallest, Adrastea, is a mere 20 kilometres (12.5 miles) across. They are all oddly shaped because they lack the mass to pull themselves into a spherical form.

SATURN:
Probing the Ringed Planet

Nearly 1.3 billion kilometres (800 million miles) beyond Earth's orbit spins a planet with a set of unique and beautiful companions – shimmering, continuously changing rings that spread for tens of thousands of kilometres above the planet itself, against the black backdrop of space.

Welcome to Saturn! Here is a planet that is distinctly different from tempestuous Jupiter, even when one disregards those circles of sunlit ice, dust and rock fragments. While Pioneer 11 was the first spacecraft to reach Saturn, a decade elapsed before the two Voyagers revealed the breathtaking majesty of this gorgeous, hazy planet with pastel-coloured clouds that are whipped into frothy bands by easterly winds of nearly 400 metres per second (900 mph). These storms were seen in detail by the Hubble telescope in 1990, before corrections were made to its optical system. Then, in 1994, the Wiffpick camera picked up the image of a huge 'thunderhead' cloud of ammonia ice, nearly 13,000 kilometres (8,000 miles) across.

The vast catalogue of Voyager pictures included our first views of all of the planet's major satellites except Titan, which is shrouded by orange hazes suspended in a 90 per cent nitrogen atmosphere. Added to that was amazing detail of the famous rings, which, though delicately thin, cover an area approximately 80 times that of Earth, and are bright enough to reflect pale sunlight onto the planet's face at night.

Saturn's major rings have now been mapped; additionally the planet has many hundreds of small rings. 'The rings consist of an icy cast of trillions that march around their captor in a vast sheet of unbelievable expanse and thinness,' runs a JPL description. 'The patterns formed within this sea of ice are both simple and complex. There are circular rings, eccentric rings, kinky rings, clumpy rings, dense rings, and gossamer rings. There are ringlets and gaplets. There are resonances, spiral density waves, spiral corrugations in the otherwise flat ring plane, spokes, shepherding moons, and almost certainly unseen moonlets orbiting within the rings.'

While Galileo contributed much to our early knowledge of the closer planets, the nature of Saturn left him quite confused, and the rings were the cause. When he trained his telescope on Saturn, he saw three disk-like images in a row. What's more, these images would regularly grow in size, then diminish. This mirage-like effect was so bizarre that for a time Galileo and his telescopes were regarded in a very dubious light.

Left: Voyager 2's view of Saturn's rings from 2.7 million kilometres (1.7 million miles) out. The rings predominantly consist of chunks of ice, or ice-covered rock, ranging in size from about 1 cm (two fifths of an inch) to 5 metres (16.5 feet). Saturn's rings are incredibly thin, certainly no thicker than 200 metres (650 feet) and perhaps as little as 10 metres (33 feet).

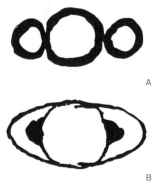

A

B

Above: Two sketches of Saturn made by Galileo in 1610 (A) and 1616 (B). In 1610 his telescope was incapable of resolving the rings adequately and he interpreted the view as a planet closely accompanied by two satellites. By 1616 his equipment had improved and his observations were more accurate.

Like Earth's, Saturn's equator is tilted relative to its equator. As Saturn moves along its 27-year orbit, first one hemisphere, then the other is tilted towards the Sun. This cyclical change causes seasons and a magnificent display of Saturn's rings. These Hubble Space Telescope images, captured between 1996 and 2000, show Saturn as its northern hemisphere moves from autumn into winter.

Eventually a telescope cleared up the mystery, thanks to the efforts of that remarkable Dutchman, Christiaan Huygens, who brought humankind the pendulum clock, the wave theory of light – and the rings of Saturn.

Galileo had already seen the rings, but he had not recognized them as such. Instead he had seen the disk of Saturn and the two bright outer edges of the rings, which had been distorted into round blobs of light.

Fourteen years after Galileo's death, in 1656, Huygens turned from his telescope and, unsure that he was correct, wrote a strange anagram to record his discovery secretly, lest he be ridiculed by his peers. Another three years of study passed before he would translate its meaning: 'Saturn is surrounded by a thin flat ring inclined to the ecliptic and nowhere touching the planet.'

Huygens' announcement must have been noted in Italy by Cassini, who – as head of the new Paris Observatory – was to discover four Saturnian satellites and collect valuable new information on the rings.

As Cassini studied the spectacular image of Saturn, he found that it possessed not a single ring, but at least two concentric rings. He announced this discovery in 1685, and Cassini's Division – the space between the two brightest rings – became part of the astronomy books. The division is about 3,500 kilometres (2,200 miles) wide and probably formed as the gravitational fields of the satellites Mimas and Janus collected ring material from what otherwise would be an uninterrupted disk. Later, a second division was detected within the outer ring, and named after the German astronomer Johann Franz Encke (1791–1865).

Saturn's wondrous rings were chosen for special attention by the Cassini spacecraft. What is their composition and how were they formed? How do they

interact with each other and their neighbouring moons? For decades we've believed that they are ice-covered fragments of rock, saved by some unknown phenomenon from smashing each other to bits. But the information is both incomplete and tantalizing. It will take more observations from the Cassini spacecraft to slake our thirst for knowledge of those silver belts.

Inside Saturn

Saturn's disk, which sometimes appears yellowish from Earth, has a pronounced oblate (flattened) shape, its equatorial girth being almost 10 per cent greater than its circumference around the poles. Like Jupiter, Saturn is gaseous and enormous; but it has the lowest density of any planet in the solar system. At only 0.69 gram per cubic centimetre (42 lb per cubic foot), an average chunk of Saturn would easily float in water.

About 86 per cent of Saturn's atmosphere is hydrogen, laced largely with helium. The atmosphere, which includes a small long-lived red spot in the southern hemisphere, also contains small amounts of methane, ammonia, ethane, acetylene, and phosphine. The colder temperatures at the top of Saturn's atmosphere allow clouds of ammonia ice crystals to condense there, forming a creamy haze that mutes the colours below. The planet's nucleus is believed to consist of two concentric layers – an outer region of molecular hydrogen and helium, and an inner one of metallic liquid hydrogen – surrounding a core of molten ice and rock. This model is similar to the one proposed for Jupiter, except that it is smaller and the various layers have different thicknesses due to gravitational effects.

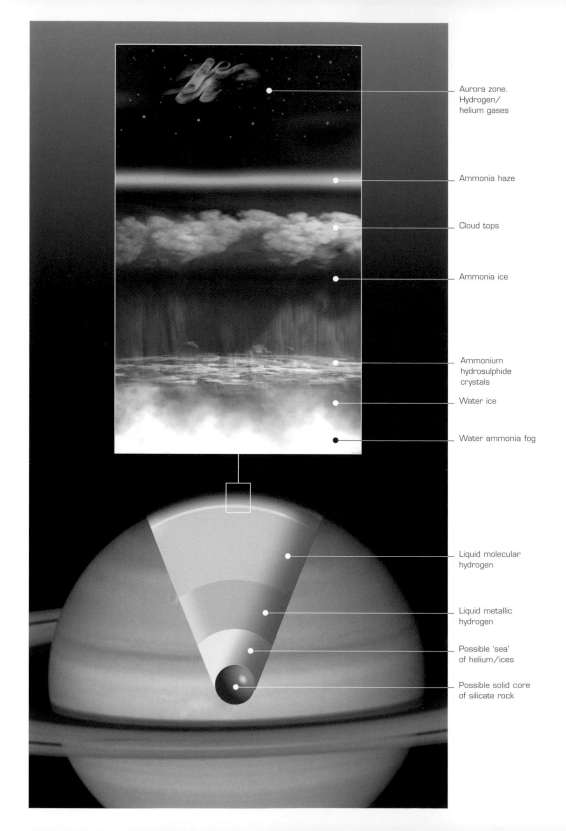

Aurora zone.
Hydrogen/
helium gases

Ammonia haze

Cloud tops

Ammonia ice

Ammonium
hydrosulphide
crystals

Water ice

Water ammonia fog

Liquid molecular
hydrogen

Liquid metallic
hydrogen

Possible 'sea'
of helium/ices

Possible solid core
of silicate rock

Also like Jupiter, Saturn proved to be extremely dynamic, radiating about 80 per cent more energy than it receives from the Sun. Part of this dynamism was evident when a storm 40,000 miles long – nearly twice the distance around Earth – was observed by instruments trained on Saturn's cloud tops from Voyager 1.

This probe also confirmed Pioneer 11's discovery of a magnetic field around Saturn, picking up periodic radio bursts from the field's rotation in January 1980. While other scientists made the invisible magnetosphere of Jupiter seem more real by producing computer graphics from traces of field lines, scientist Frederick L. Scarf painted his pictures with sound. Scarf, a physicist from the TRW corporation working on Voyager's plasma wave experiments, noted that radio waves at acoustic frequencies came from the plasma that formed around the spacecraft as it moved through Saturn's magnetosphere. So he fed the signal through an ordinary loudspeaker. The result was pleasing enough that Scarf and friends played their data through a 16-channel music synthesizer as Voyager 2 drove through the Saturnian magnetic field.

Secrets of Light

Identification of ammonia and methane in Saturn's atmosphere followed the analysis of red and infrared radiation in the planet's reflected light. First suggested by Rupert Wildt of the University of Göttingen in 1932, this discovery was confirmed shortly afterward by Theodore Dunham at Mount Wilson Observatory in California.

Use of the spectrum to analyse solar and planetary chemistry originated in the nineteenth century, when scientists were studying the rainbow of colours produced from white light by a prism. Though many scientists were involved in this early work, the most famous was Joseph von Fraunhofer, the German physicist who studied the light emitted by stars and reflected from planets. Fraunhofer showed that spectra contain or lack certain colours (wavelengths of light) that are emitted or absorbed by certain chemical elements. Thus, if the light reflected by a given object fails to reflect the characteristic wavelength of ammonia (as is the case with Saturn), then it follows that ammonia is part of its atmosphere.

When we look at planetary spectra, we learn which elements are present in that part of the atmosphere that reflects or absorbs light.

Light emitted or reflected by certain objects can be used to produce images, much as night vision equipment produces images of the infrared wavelengths

Left: A theoretical cross-section of Saturn.

The appearance of Titan's dense atmosphere changes according to the seasons. This Voyager 2 image shows the northern hemisphere leaving winter. It is noticeably darker than the southern hemisphere, but this effect will reverse as the seasons progress.

given off as body heat by living creatures. Spacecraft cameras are equipped with filters so that they also have the equivalent of night vision for various chemicals. Different filters let through wavelengths of light that, when analysed individually or in combination, signal the presence and abundance of these chemicals. Cassini's wide angle camera is equipped with 18 filters, its narrow angle camera with 24 filters.

Not surprisingly, the Imaging Science Subsystem was designed to be one of the chief scientists of Cassini's robotic crew. It was charged to learn as much as possible about the atmospheres of Saturn and Titan; to map them in 3D; to investigate their clouds; and to look for lightning and aurorae. It was also expected to glean more knowledge of the rings' structure and dynamics, and the way in which they interact with nearby moons. And the satellites themselves? The Imaging Science Subsystem was programmed to map their surfaces in detail and find clues to their geological histories.

Saturnian Satellites

The rings of Saturn are marvellous, but its larger moons, like those of Jupiter, are among the most exciting places we've found beyond the asteroid belt. The more we look, the more moons we see circling this great planet. Late in the year 2000, a group of astronomers from France, Canada, and the U.S. discovered twelve new Saturnian satellites, bringing the total to thirty.

Saturn's lunar giant is Titan, larger than the planets Mercury and Pluto. The second largest moon in the solar system after Ganymede, it measures 5,140 kilometres (3,200 miles) across. It has an atmosphere composed almost entirely of nitrogen, mixed with methane and other hydrocarbons, and it possesses an orange haze so thick that white light cannot penetrate it.

It's thought likely that Titan's surface is partially covered with liquid or sludgy hydrocarbons such as ethane, that evaporate into smog and then condense as rain, in a process that mimics Earth's own water cycle. Titan's atmosphere also supports a mild greenhouse effect, and the environment may be similar to that which existed on our planet before the emergence of life. In

fact, chemical reactions now occurring on Titan could produce the molecules that are believed to have sparked the beginning of life on Earth.

'There are only three objects – Mars, Europa, Titan – that may have undergone significant organic chemical evolution,' says Jonathan Lunine, an ESA Huygens researcher at the Lunar and Planetary Observatory, University of Arizona. 'This evolution may have progressed even to the threshold of life, perhaps, on Titan – possibly beyond it on Mars and maybe Europa.

'Titan may host the kinds of organic chemical reactions that preceded and initiated life on Earth 4 billion years ago. To see methane in action, forming clouds and rain, makes Titan a very attractive astrobiology target.'

Dominic Fortes of University College London envisages an exotic biosphere existing in an ocean of ammonia-water solution as much as 200 kilometres (125 miles) deep, under a water-ice crust. 'Conditions,' he wrote in the planetary sciences journal *Icarus*, 'are such that life could well exist and that the atmospheric methane and nitrogen may be of biogenic origin.'

In August 2000, observers from Paris-Meudon Observatory announced that they had found bright reflecting areas in Titan's midsection. Infrared images, captured with the 3.6-m (12-ft) Canada-France-Hawaii Telescope using adaptive optics, may show methane ice, formed atop a mountainous plateau. Three months later, astronomers at Northern Arizona University discovered that sparse clouds appear in Titan's atmosphere each day. The plot thickens.

Layers of smog-like haze covering Titan can be seen in this image taken by Voyager 1. Two layers of haze (coloured blue) can be seen above the main atmosphere (coloured orange). The divisions in the haze occur at altitudes of 200, 375 and 500 kilometers (124, 233 and 310 miles) above the surface of the moon.

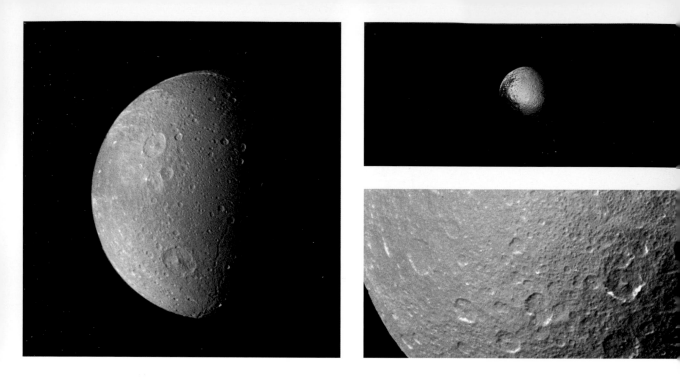

A gallery of five of Saturn's larger moons. From left to right: Dione; Iapetus (top); Rhea (bottom); Tethys; Enceladus.

ESA's Huygens probe is designed to look beneath Titan's clouds for surface liquids, while other experiments on the Cassini orbiter and the probe investigate the chemical processes occurring in this strange atmosphere. Scientists hope that this investigation, of a body that may be covered with lakes or oceans of methane or ethane – or even a water/ammonia antifreeze mix – will 'provide the equivalent of a time machine to examine the chemistry of the early Earth.'

With its photochemical haze (on Earth, we'd call it a smog), Titan was the only major moon whose surface remained hidden from the Voyager spacecraft. The exposed surfaces of Tethys, Dione, Rhea, and Iapetus – relatively large moons discovered by Cassini in the seventeenth century – were imaged, as were those of Enceladus, Mimas, Hyperion, Phoebe, and eight other satellites.

Mimas, the closest larger moon to Saturn, has a huge crater, about 80 miles (130 kilometres) across, named for the moon's discoverer, Sir William Herschel. It also has several smaller craters, along with a mysterious series of grooves. Caused by a gigantic meteor impact, Herschel is about a third the diameter of the entire moon, making Mimas look rather like a huge navel orange.

The unusual and varied texture of Enceladus suggests that at one time it may have been as active as Jupiter's Io. Parts of its surface bear the cold remains of volcanic flows that could have been generated by the welling up of interior heat, combined with the huge tidal pull of Saturn's mass. Enceladus is the

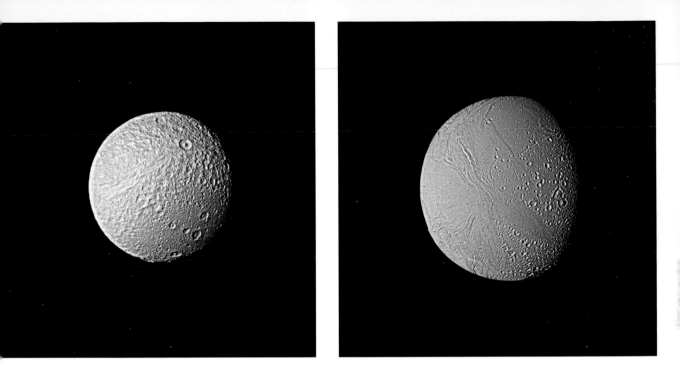

most reflective satellite in the solar system, and this may indicate that ice and snow 'geysers' are still active on its surface.

Tethys and Dione are nearly the same size, both about 1100 kilometres (700 miles) across. Dione is a frost-scarred sphere of ice and rock, and Tethys a ball of ice marked with numerous bowl-shaped craters and a huge valley some 800 kilometres (500 miles) long. Rhea, the biggest satellite after Titan, is a somewhat larger version of Tethys.

Beyond the orbit of Titan lies Iapetus, another ice ball 1,440 kilometres (895 miles) across. Iapetus is unusual in that one side is dark coloured, perhaps with organic material, while the other is pallid. The moon's 'black eye' may be some kind of volcanic artefact. But researchers using the U.K. Infrared Telescope in Hawaii think it may be sooty organic matter, possibly sprayed out from a giant meteorite impact on Titan.

Scientists examining Voyager 2 images reasoned that the most distant large Jovian moon, Phoebe, because of its dark surface and unusual orbit ('backwards' from Saturn's spin), could be a captured asteroid. It is another possible source of Iapetus' dark coating.

In the last three months of 2000, twelve new moons were found orbiting Saturn, bringing the total to thirty. Some were only a few kilometres in diameter; they may have originated as Centaurs, the icy bodies that orbit between Saturn and Uranus.

Ah, Sweet Mystery ...

How life first arose on Earth is one of the great mysteries of science. Christopher Dobson of Oxford University, working with Adrian Tuck of the U.S. National Oceanic and Atmospheric Administration (NOAA) and Barney Ellison and Veronica Vaida of the University of Colorado, suggest that it could have originated in tiny airborne bubbles – aerosols. Organic materials within them would have concentrated with time, and been exposed to all manner of radiation, humidity, and temperature changes. Dobson and his colleagues suggest that this would be an ideal condition for pre-biotic chemicals to self-assemble and begin to morph into living protoplasm.

Life could also have come here from somewhere else. Pre-biotic molecules might have migrated here as hitch-hikers on comets, asteroids, or interstellar dust. These, and possibly more complex forms, might have arrived on Earth by way of meteorites.

Structures resembling bacteria were found in the famous meteorite ALH84001. This is a chunk of Martian rock – we know its origins because it contains traces of gas identical to the Martian atmosphere analysed by the Viking landers – that is 3.6 billion years old. Could a meteorite such as this bring life to Earth? Many scientists are dubious that this particular rock actually contains a fossil, yet studies of its magnetism suggest that it never got hotter than 40°C (104°F) during its journey from Mars to Earth. That's not hot enough to destroy bacteria or fungi or even plant seeds.

But chunks of Mars are flung onto their flight paths toward Earth after they are blasted from the surface by massive meteorite impacts. Could bacteria survive this enormous shock? Even this question has been investigated. A German team subjected a culture of bacterial spores to a simulated impact – and enough survived at least to start a long journey Earthward, 'in a scenario of interplanetary transfer of life.'

Examination of ALH84001 by a number of government, academic, and independent scientists in 1996 came to the conclusion that the hydrocarbons, carbonates, and general structure of this meteorite could be 'fossil remains of a past martian biota.' Four years later, the same group teamed with colleagues to produce more evidence of possible past life from the same meteorite – magnetite crystals similar to those produced by modern bacteria. These, the scientists suggested, 'were likely produced by a biogenic process': the crystals were 'interpreted as martian magnetofossils.' Almost at the same time, another team of scientists poured cold water on the idea, describing a non-biological mechanism for producing these materials – 'a simple inorganic process for formation of carbonates, magnetite, and sulphides in Martian meteorite ALH84001.'

Earthbound fossil records, recovered from ancient sediments excavated in deep drilling operations, indicate that life first appeared on our planet about 3.8 billion years ago, when Earth's crust had newly settled and the early atmosphere and oceans were established,

For as much as three billion years, all life consisted of single-celled micro-organisms, Then multi-celled organisms began to appear, multiplying rapidly in the so-called Cambrian Transition. Because the transition occurred so quickly in geological terms – a few million years – it is sometimes called the Cambrian Explosion,

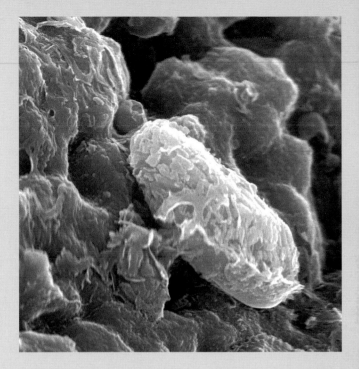

This cocoon-shaped object, found on the surface of a meteorite discovered in Nakhala, India, may be a fossilized Martian cell. It is embedded in a clay mineral known to have formed on Mars about 700 million years ago.

About 250 million years ago, a global catastrophe killed off an estimated 70 per cent of land vertebrates and 90 per cent of marine species. Apparently this mass extinction was caused by an asteroid impact enormous enough to spew more than a million cubic kilometres (a quarter million cubic miles) of dust into the atmosphere – sufficient to reduce Earth's temperature drastically, decimating plant life and many of the animal species that depended upon it. Life, of course, survived. Eventually the dinosaurs arose and thrived, only to be extinguished in another massive climatic disaster that paved the way for the rise of mammals (see p.71).

But many ancient members of the living micro-world are still with us, apparently inextinguishable, unaffected by the mass extinctions. The variety, and the amazing toughness, of some of these micro-organisms, has only recently been appreciated. Their existence heightens the expectation that life exists on other planets.

Earth creatures in general require liquid water and the right mix of chemicals in order to sustain life, with sunlight providing the basic energy source. Bacteria exist today that can salvage the solar radiation that manages to penetrate through several metres of solid ice. There are, moreover, a relatively small number of strange organisms, neither plant nor animal, that

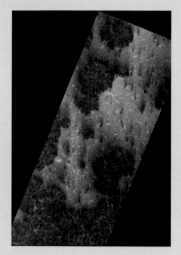

This strip of data from the Mars Global Surveyor includes life-like shapes that have been compared with banyan trees. But scientists provide a geological explanation for the shapes, which in some ways resemble weathered features on Venus.

derive their energy from other means. Some thrive upon reactions that are fired by the heat of geothermal energy, their metabolism dependent upon a supply of oxidants (including sulphur dioxide and carbon dioxide) that have migrated downward from the surface. These creatures live around the hot vents and volcanic fissures that dot the tectonic cracks in the ocean floors, and inhabit heated aquifers kilometres below Earth's surface.

Other organisms are at home with high levels of radiation. *Deinococcus radiodurans* was discovered in 1956 when scientists were testing ways of sterilizing meat. When the meat was exposed to radiation, every other organism perished except this one. Because it is able to repair its own DNA, it can withstand well over a thousand times the radiation dosage that is fatal to humans, and it is drought-resistant. Perhaps *Deinococcus radiodurans* originally came from outer space.

Still other bacteria have proven to live virtually forever. One previously unknown bacillus, similar to organisms found in the modern Dead Sea, was found in an ancient salt crystal, apparently in a state of suspended animation, and brought back to life. This amazing scrap of organic matter was found in an air intake shaft at a salt mine used for storage of low-level radioactive waste in New Mexico, part of a 250-million-year-old formation located more than 600 metres (1000 ft) underground.

How significant are these discoveries in the scheme of things? On March 22, 1998, some boys found a stony meteorite that had landed in Texas. Miraculously, the meteorite included a patch of salt crystals, and within these crystals were small droplets of water, estimated at as much as 4.5 billion years old. Scientists puzzled over this mystery, and some wondered, what if this water had contained the sleeping spore of some exotic organism. The boys had something else in mind. They sold the meteorite on the Internet.

Such incidents, and the still-unfolding story of strange and ancient organisms – terrestrial extremophiles – illustrate that life can thrive in what we previously believed to be unearthly circumstances, and lends credence to the theory that living creatures exist elsewhere in the solar system.

The habitability of planets beyond Earth – and the possibility that organisms already live there – has always been of great interest to both armchair and professional scientists. Some scientists, of whom Carl Sagan of Cornell University was the best known, have seen distinct possibilities that some kind of life already exists on other planets within our own solar system.

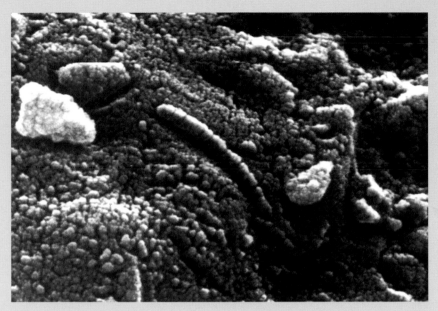

The structures seen in the meteorite ALH84001 are only 20 to 100 billionths of a meter across.

Even when it seems that planetary surfaces are too hot or too cold to sustain life, or that the chemical environment is somehow wrong, there's always the possibility that organisms unfamiliar to us may have developed within their own 'level of comfort,' much as marine organisms did in Earth's own biosphere.

Doubtless we will deliberate forever the possibilities of extraterrestrial life. Still there is wonder enough on Earth, where Charles Darwin's thoughts remain valid after nearly a century and a half. 'Probably all the organic beings which have ever lived on this earth have descended from some one primordial form,' he wrote in *The Origin of Species*. 'There is grandeur in this view of life... that, whilst this Planet has gone cycling on according to the fixed law of gravity, from so simple a beginning endless forms most beautiful and most wonderful have been, and are being, evolved.'

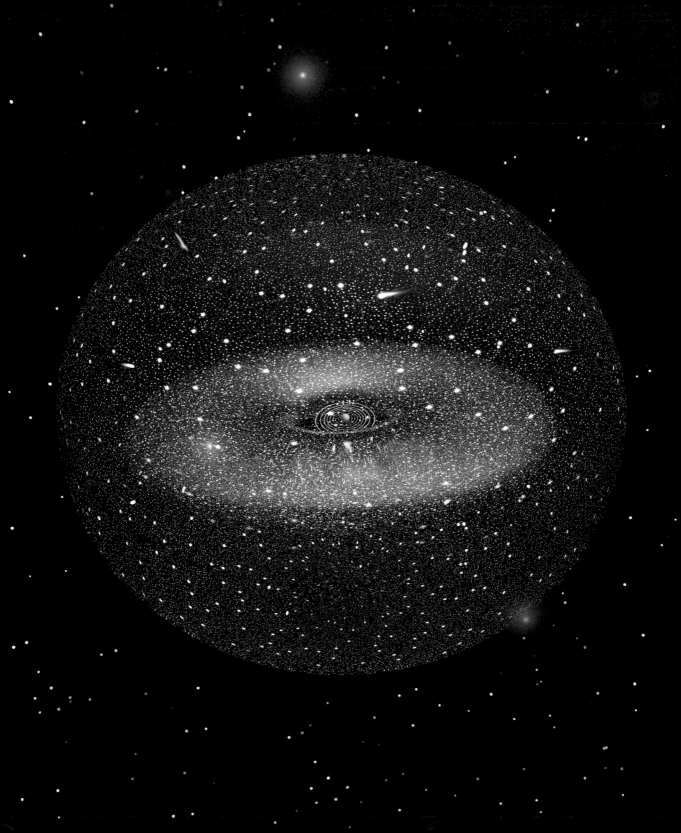

The Outer Limits

'The choice, as Wells once said, is the Universe – or nothing... The challenge of the great spaces between the worlds is a stupendous one; but if we fail to meet it, the story of our race will be drawing to its close. Humanity will have turned its back upon the still untrodden heights and will be descending again the long slope that stretches, across a thousand million years of time, down to the shores of the primeval sea.'

Arthur C. Clarke

Beyond Saturn is a cold, distant realm that has been visited by a single spacecraft and whose details can be glimpsed only dimly by the most capable astronomical instrument ever created, the Hubble Space Telescope. These are the mysterious outer limits of the solar system, beginning at Uranus, orbiting 2.9 billion kilometres (1.8 billion miles) from the Sun, and taking in Neptune, Pluto, Charon, and the Kuiper Belt.

Uranus and Neptune are giant ice planets that are similar in size, similar in density, similar in that they have oddly off-centre magnetic fields, and aurorae at their poles. But they are non-identical twins only: for example Uranus has no perceptible internal heat source, and we have yet to see aurorae around Neptune's magnetic poles. They are also different colours. Uranus is greenish blue, indicating a subtle difference in composition from deep-blue Neptune.

How did these low-density giants get out there beyond the orbit of Saturn? It now seems that both Uranus and Neptune formed much closer to the Sun, but – at some point in the early aeons of the solar system – they began to move outward from their neighbouring protoplanets, with Neptune's migration accounting for nearly a third of its present 4.5 billion kilometres (2.8 billion miles or 30 A.U.) average distance from the Sun. How and why that happened is an unanswered question, part of the vast volumes that we yet have to learn about the formation of our planetary system.

URANUS:
The Tilted Planet

Voyager 2's arrival at Uranus almost coincided with the 200th anniversary of the planet's discovery. The story began in March, 1781, when Britain's austere Royal Society received a note from a scholarly German gentleman who was known as an expert in the design of large telescopes. The note said simply that he had discovered a new comet in the constellation of Taurus.

The astronomer had noted a couple of strange things about this 'comet.' It was very well defined, and it appeared to have no tail.

No one since ancient times had discovered a planet, so the astronomer perhaps reluctantly held to the comet theory until someone else announced that the object was a new planet, occupying an orbit about 19 A.U. from the Sun. With this encouragement, he came forward and claimed his right as discoverer, astutely naming the planet George's Star in honour of King George III. Later Johann Elert Bode renamed it Uranus.

The discoverer of Uranus was none other than the Sir William Herschel we met earlier. He made the first systematic study of the brightness of stars, put together a massive catalogue of nebulae and double stars, correctly determined the rotation period of Saturn (as well as discovering its moons Enceladus and Mimas), and discovered the motion of the solar system through space.

Herschel also found Uranus's two largest satellites, Titania and Oberon, in 1781. England's William Lassell (1799–1880) discovered Ariel and Umbriel in 1851, and Gerard Kuiper found a fifth satellite, Miranda, in 1951.

Uranus is tinted gently with soft shades of greenish blue. A most perplexing planet, it circles its orbital path virtually 'on its side,' with tenuous, wobbly rings and a corkscrew magnetic field that stretches millions of kilometres away from the Sun. Most other planets spin with their rotational axes upright or nearly upright with respect to their orbital planes. Strange-natured Uranus rotates on nearly the same plane.

The most widely accepted explanation for Uranus' bizarre spin is that it may have been involved in a huge interplanetary collision sometime in its early history. To knock Uranus into such an unusual posture, the other planet must have been at least the size of Earth and perhaps even twice as large.

On its arrival in the winter of 1985/6, Voyager 2 found a cold planet with a remarkably featureless atmosphere. Examination of nearly six thousand telemetered images showed little detail even after the planet had been scanned from Voyager's closest approach of 8,000 kilometres (5,000 miles) above the

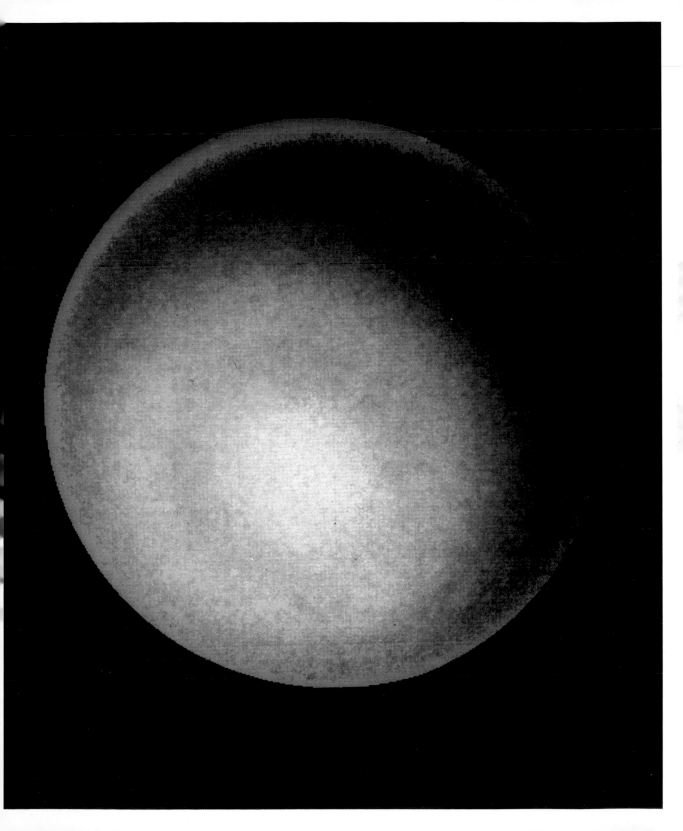

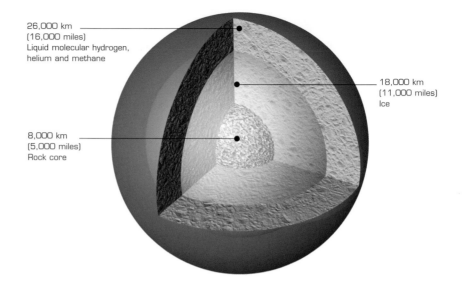

26,000 km
(16,000 miles)
Liquid molecular hydrogen,
helium and methane

18,000 km
(11,000 miles)
Ice

8,000 km
(5,000 miles)
Rock core

A possible model for the interior of Uranus. Alternative models suggest a liquid rather than a rocky core. The atmosphere of Uranus is composed of 83 per cent hydrogen, 15 per cent helium, 2 per cent methane, and trace amounts of acetylene and other hydrocarbons.

planet's visible surface. The spacecraft peered at Uranus's five known moons, and revealed the existence of ten that had never been seen before. It also discovered two new rings, bringing the total to eleven. Miranda, smallest of the large moons and closest to Uranus' surface, turned out to have the most complex surface ever seen in the solar system (see box on p.234).

But the Voyager flyby of Uranus itself left us with a wintertime picture that was about as bland as a golf ball. Uranus's year is equivalent to 84 Earth years, and for nearly a quarter of this time, the Sun shines directly onto one of the poles while the other pole is shrouded in utter darkness. The planet would have been a more exciting place later, closer to the vernal equinox of 2007 when both hemispheres will be illuminated. Fortunately, the Hubble Space Telescope is able to track Uranus's springtime weather patterns as they come to life in the quickening light of the distant Sun.

Now let's take a look below Uranus's greenish drapery of filmy methane clouds – which give the planet its colour by absorbing red light. The popular model shows us a three-layered planet – an upper atmosphere that is about 98 per cent hydrogen and helium, characterized by winds of 100–600 kilometres per hour (60–375 miles an hour) in the illuminated southern latitudes; an icy mantle containing water, methane and ammonia; and a dense core. The discovery of helium and hydrogen on this planet was no surprise – it makes sense that all the giant planets must contain a large percentage of the light elements, for these were the principal constituents of the protostellar disc. Theory suggests that the cores of Uranus and Neptune are composed of molten rock, ice, and gas from the solar system's first asteroids and comets.

The hydrogen/helium ratio of Uranus's atmosphere – and indeed at all the giant outer planets – was measured by the Voyagers with instruments that measure the intensities of radiation within the infrared (IR) part of the spectrum. IR spectrometers measure several frequency ranges; IR photometers measure just one range.

The Voyagers each flew with an Infrared Interferometer Spectrometer and Radiometer (IRIS). This high-tech light meter included a telescope through which light was separated into two beams, then filtered and directed to a crystal detector for measurement of its intensity. As needed, it could act as a sophisticated thermometer, as a detector of certain elements or compounds, or (with a separate radiometer) as a way of measuring the sunlight reflected at UV, visible, and IR frequencies. It was also able to measure the temperatures at different altitudes and to determine whether planets are radiating more energy than they receive.

From absorption and emission spectra, these instruments measured concentrations of hydrogen, helium, water, methane, acetylene, ethane, and ammonia in the atmospheres of all four giant planets.

Little or no heat comes from the interior of Uranus; indeed, unlike Jupiter, Saturn and Neptune, it may give off no more energy than it receives from the Sun. It may never have contained a heat source; that heat source may have run out of energy; or there may still be one deep within the planet, somehow contained by the atmosphere.

Uranus's magnetic field is peculiar in that it is tilted by 60 degrees from the planet's rotational axis. It produces aurorae with a brightness of about 3 billion watts, which is 10 times more that of Earth and probably the greatest in the solar system. How are they generated? Floyd Herbert of the University of Arizona suggests that ions (rather than the more usual iron) may be responsible. The magnetic field may be generated closer to the surface at Uranus and Neptune than on Earth, 'perhaps in a subsurface layer where the chemistry favours molecular ions.' This layer may also be the reason for the off-centre position of the planet's magnetic field.

The orbits of Uranus' five largest satellites – which lie in the same plane as the planet's equator – vary so much with respect to Earth that through telescopes we sometimes see them edge-on and sometimes almost face-on. These moons bear the names Miranda, Ariel, Umbriel, Titania, and Oberon,

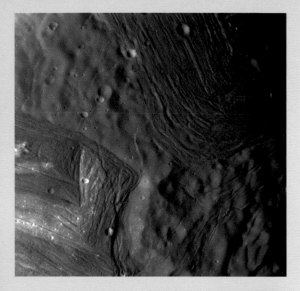

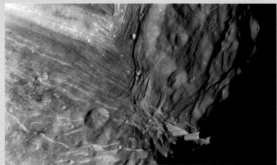

Miranda

Though it is only about one-sixth the size of Earth's moon, Miranda caused a sensation, for it is surely one of the strangest bodies ever observed in the solar system. It looks as though a regular spherical world, pitted with Moon-like craters, had been given a series of glancing blows with a huge sledgehammer, leaving giant scrapes and gouges on its surface. On closer inspection, the 'gouges' turn out to have a fine structure of concentric canyons, one within the other. Some of these are nearly 20 kilometres (12 miles) deep, and are marked with a smaller concentration of craters than the surrounding terrain.

How did Miranda acquire its bizarre topography? One theory of several is that, very early in the history of the solar system, gargantuan blocks of ice floated on the moon's molten spherical bed. These gradually settled into a fairly even, spherical shape. Then the temperature cooled, and three immense blocks were left frozen in the exposed surface, edged by petrified ripples of planetary material. While the surface has been modified somewhat since then, it retains its peculiar patterns and shapes.

We only have such good photographs of Miranda because the Voyager 2 spacecraft had to skim close by in order to get the gravity boost necessary to send it onward to Neptune. Images of this moon were made in January 1986, from a distance of 147,000 kilometres (91,000 miles).

Voyager skimmed so close to the surface that it was possible to make a remarkable short video of the encounter. The images started with a series of photos compiled by the U.S. Geophysical Survey at Flagstaff, Arizona. These were made into a mosaic and processed into a continuous series of changing, 3D-like views by a computer technique called photoclinometry. Finally, with the addition of a little music, JPL's Helen Mortensen, Kevin Hussey, and Jeff Hall produced 'Miranda the Movie,' a simulated fly-over constructed from real data.

If planetary scientists had their choice when planning the tour of Uranus, they might have foregone a Miranda flyby in favour of visiting Oberon and Titania. In retrospect, said JPL's Torrence Johnson, 'we were very fortunate that God and Kepler more or less dictated that we had to go by Miranda.'

Top: This image of Miranda, obtained by Voyager 2 on approach, shows an unusual 'chevron' figure and regions of distinctly differing terrain that bear witness to the Uranian moon's complex geologic history.

Above: Two distinct terrain types characterize Miranda: rugged, higher-elevation terrain on the right and lower, striated terrain on the left. Numerous craters on the rugged, higher terrain indicate that it is older than the lower terrain. Several fault scarps cut across the different terrains.

Right: A south polar view of Miranda assembled from Voyager 2 images.

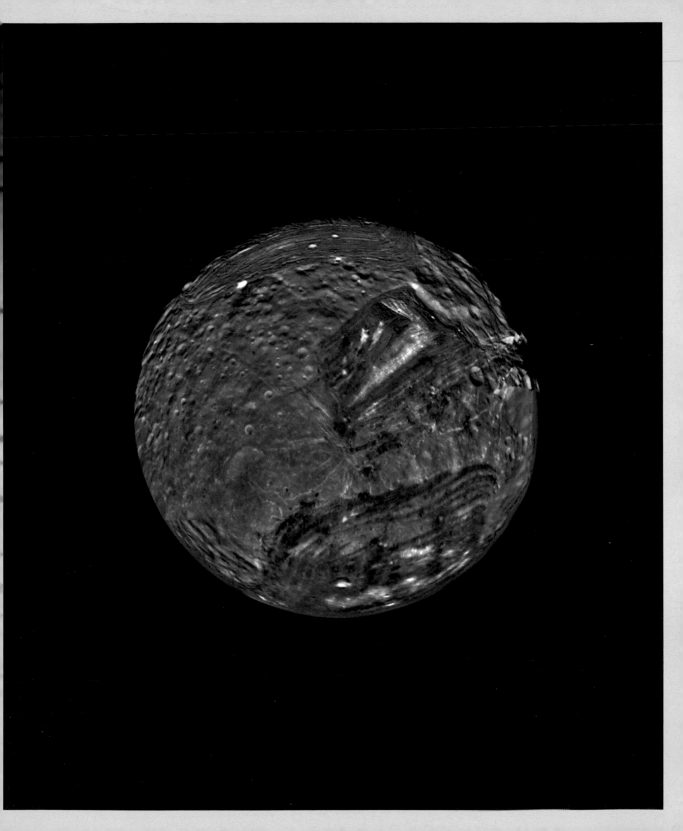

This Hubble Space Telescope image of Uranus captured not only its rings and some of its satellites, but also revealed several bright clouds in the atmosphere.

taken from the works of Shakespeare and Pope. Voyager 2 found that they had at least ten companions, named in like fashion Puck, Belinda, Cressida, Portia, Rosalind, Desdemona, Juliet, Bianca, Ophelia, and Cordelia. Caliban and Sycorax were discovered in 1997 by Philip Nicholson of Cornell University, and Erich Karkoschka of the University of Arizona found another moon by comparing imagery from Hubble to that of Voyager 2. Thus our changing picture of Uranus ended the old century with a total of 18 verified moons – since when a further six faint, irregular moons (probably captured asteroids) have been sighted with the Canada-France-Hawaii Telescope (CFHT) on Mauna Kea, and new names have joined the fleet of circling moons – Prospero, Setebos, Stephano.

Of the major moons, four have been 'resurfaced' by geologic activity that from time to time erases some of the traces of meteor impacts. These moons are not, as was thought in pre-Voyager times, inert and ancient orbiting globes of ice and rock – only Umbriel matches that description, having experienced neither the internal heat nor the external tidal forces that have affected the other moons. All are comparable in density with the icy moons of Saturn, at more than 1.5 grams per cubic centimetre (89 lb per cubic foot).

The first six Uranian rings were discovered from the airborne Kuiper Observatory when Uranus passed in front of a star in 1977. Perth Observatory found another three, then Voyager 2 found two more. All are dark and colourless, ranging from 1 to 100 kilometres (half a mile to 60 miles) wide and measuring only a kilometre or two thick.

It would seem that these rings should 'want' to spread out – continuing collisions among their billions of particles would normally give them a kind of runniness. But they don't spread out. No one knows why they stay the same width, but it could be a result of the 'shepherding' influence of nearby moons, sweeping up rogue particles.

Voyager 2 was the first spacecraft to visit Uranus, and is likely to be the only one to venture there for decades to come. Having provided its Earthbound masters with new information on yet another poorly understood planet, it set forth for Neptune in the early spring of 1986.

In a sense we have left the best part of the Uranus story until last, for users of the Hubble Space Telescope are having the luxury of watching this planet awaken with spring. Heidi Hammel felt that Uranus had been the underdog of the solar system. 'I call it the Rodney Dangerfield of the outer planets,' she says. 'It doesn't get any respect.' But now, thanks to the new observations being made with the Hubble telescope, 'we're going to really change people's ideas of this planet.'

Interest in Uranus was stirred in 1994, when the Hubble Space Telescope's Wiffpick camera imaged surprisingly high-contrast clouds on the planet's disc. Hammel and her fellow astronomers could now pick out clear conditions in some parts of the atmosphere and hazy regions around the poles. Clouds the size of terrestrial continents were seen on the planet's northern hemisphere, reaching high into the atmosphere. The rings were visible, especially so in near-infrared wavelengths, as were eight of the ten satellites discovered previously by Voyager 2.

In 1998, Hubble observers found 20 clouds, some moving at more than 500 kilometres per hour (310 mph). Clearly the approaching spring was changing the face of Uranus. By equinox the planet may develop a planetary personality as lively as Jupiter. But we'll have to wait for that.

NEPTUNE:
Rendezvous with a Blue World

It seems remarkable that in 1840, when we already had railway locomotives and early electric motors, no one knew that Neptune even existed. We'd known about Uranus for more than half a century, and to all intents and purposes it was the outermost sentinel of the solar system. But there were problems. Uranus just didn't behave in a predictable way.

Planet watchers observed Uranus quite contentedly until the turn of the nineteenth century. But then they noticed it didn't move as predicted along its orbital path; indeed it seemed to lag behind. What was causing the planet's mysterious behaviour?

In three countries three mathematicians decided independently to find out. Their names were Friedrich W. Bessel (1784–1846) of Germany; Urbain J. J. LeVerrier (1811–1877) of France; and John Couch Adams (1819–1892) of England. All three concluded that an unidentified planet was pulling at Uranus with its gravitational field. Now their job was to predict where the invisible planet would be found.

Heidi B. Hammel
Senior Research Scientist
Space Science Institute

Exploration of Ice Giants

Beyond the realm of huge Jupiter and past the ringed world Saturn, two more giant planets circle the Sun in the darkness of deep space. They are Uranus and Neptune, somewhat smaller than their closer counterparts yet still many times more massive than Earth. Their great distance from Earth still confounds all but the world's very best telescopes. Much of what we know about these planets came from the single spacecraft that has visited both, Voyager 2.

Uranus is distinguished from the three other giant planets in our solar system by two facts. First, sometime in Uranus' past, a huge collision devastated the nascent planet. As a result, the rotation pole of Uranus is now tilted more than 90 degrees from the plane of the planet's orbit. I watched in 1994 as a small wayward comet struck Jupiter; the ensuing celestial fireworks were documented by me and observers all over Earth. Yet within just a month, Jupiter's ugly bruises had faded, and the planet spun serenely on as if nothing untoward had happened. Thus, I can only imagine what happened to poor Uranus, completely disrupting the planet to the point of twisting it completely off its normal spin axis.

Second, Uranus lacks excess heat radiating from its interior. In comparison, the other three giant planets radiate significant excess heat. The heat is thought to be left over from the time of the planets' formation and from continuing gravitational contraction. Why then does Uranus have none? Scientists theorize that perhaps the event that knocked Uranus over on its side somehow caused much of the heat to be released early in the planet's history, or perhaps the heat is there but is trapped by layers in the atmosphere.

Uranus has the most extreme seasonal variation in the solar system, thanks to its tilt. It is now approaching equinox, that time when its equator is pointing toward the Sun, and hence also our telescopes here at Earth. My colleagues and I are watching Uranus carefully, in the hopes that answers to some of our questions might be revealed as we probe the planet's seasonal change with modern astronomical instrumentation. It already appears that the northern hemisphere differs from the southern in terms of overall haziness and number of bright cloud features. With each visit to the telescope, we hold our breath just a little, because we are not quite sure exactly how the planet will look.

Neptune is the master of trickery at the telescope. Every single image holds a completely different view of this slippery planet, though you need a very big or very good telescope. Its ever-changing face, sometimes speckled with white, other times daubed with dots of darkness, never fails to stimulate my imagination. We know of the details for only one time: 1989, when Voyager 2 flew over the Neptune cloud tops. I was at one of the big telescopes in Hawaii as that was happening, documenting Neptune's ground-based appearance. By comparing these images with our Voyager views, we have a 'key' to allow us to speculate what the post-Voyager and pre-Voyager Neptune must look like at 'Voyager-scale' resolution. The most surprising thing is how very different Neptune seems to be each time we look.

Voyager detected numerous cloud features in Neptune's visible atmosphere. The biggest was the Great Dark Spot, a hurricane-like storm that was about half the size of the Earth. Yet less than five years later, Hubble Space Telescope images showed that the Great Dark Spot discovered by Voyager had disappeared, and a new Great Dark Spot had developed in the northern hemisphere. Within a few years, that feature had also faded. Neptune today is the brightest than it has ever been in 30 years of monitoring, yet we do not understood why, nor how long the brightening will continue.

On Earth, energy from the Sun drives the weather systems. However, the mechanism must be very different on Neptune, because the planet radiates almost three times more energy than it receives from the distant Sun. The rapid dynamics of Neptune's atmosphere may be caused by this strong internal heat source warming the cloud layers from below. A slight change in the temperature differential from cloud bottom to top might trigger large-scale changes in atmospheric circulation.

In their 1989 *Atlas of Uranus*, Garry Hunt and Patrick Moore commented that 'despite the space-probes, the planets guard their secrets well.' Shrouded in distant darkness, Uranus and Neptune reveal their secrets with tantalizing slowness. Nevertheless, the veil is lifting, and we are beginning to appreciate these distant ice giants as worlds of beauty and dynamic change.

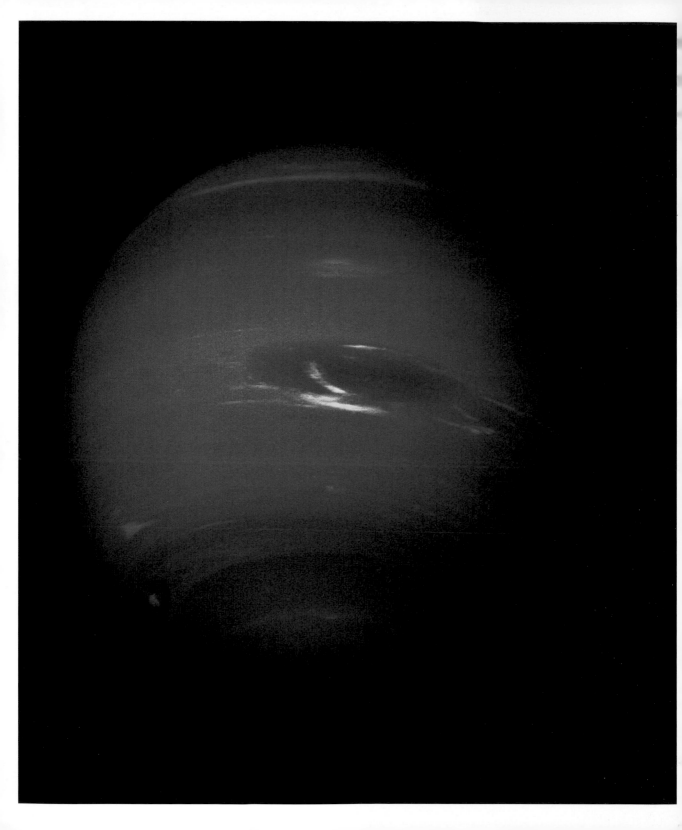

At the age of twenty-six, after two years' study following his graduation in 1843 from Cambridge, Adams decided upon the likely position of Neptune and asked the Astronomer Royal, Sir George Airy (1801–1892), to locate it visually. Airy was alternately apathetic and sceptical – until he heard that LeVerrier was predicting the same position in France.

LeVerrier apparently was some months later than Adams in finishing his calculations. But he was fortunate in that he had found an enthusiastic listener. It was Johann Galle of the Berlin Observatory, who quickly found Neptune from LeVerrier's description on September 23, 1846. Amateur astronomer William Lassell located Neptune's giant satellite Triton less than a month later.

Bad feelings surrounded the question of Neptune's discovery for many years following. Bessel's death had put him out of the running, but 'political' feelings in London almost led to the discrediting of Adams as Airy and J. C. Challis (1803–1882) of Cambridge attempted to garner England's share of the glory for themselves.

All turned out for the best, however, and the shy mathematician from Cornwall is today given joint credit with LeVerrier for the prediction of Neptune's position. Neptune is recognized as the only planet to have been located by mathematical prediction, rather than by systematic visual studies of the sky.

As with Uranus, it is now believed that Neptune once orbited the Sun much closer in, along with large numbers of planetesimals, but that it scattered most of them away through gravitational effects while collecting the remainder. As the process continued, Neptune moved outward and farther from the Sun while Jupiter hurled the scattered planetesimals out of the solar system and drifted inward.

Neptune's clouds came to light with the new technical capabilities that became available to astronomers in the closing decades of the twentieth century. In the summer of 1989, astronomer Heidi Hammel announced that she had detected clouds in visible-light images of Neptune obtained with the University of Hawaii's 2.2-metre (88-inch) telescope on Mauna Kea. Her paper, published in *Science*, predicted that the images would be of special importance for the upcoming Voyager 2 encounter with Neptune, when images taken through the spacecraft's methane band filter 'should show distinct cloud features and a polar haze.'

She was right. On August 25, 1989, Voyager 2 sailed within 5,000 kilometres (3,000 miles) of the pale blue cloud tops near Neptune's north pole. Five hours

Voyager 2 revealed Neptune to be a dynamic planet with hurricane-like storms reminiscent of Jupiter's Red Spot. One of these storms, dubbed The Great Dark Spot, was the size of Earth and can be seen in the centre of this image.

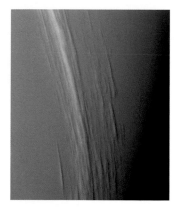

Bands of high-altitude cirrus clouds cast shadows on the atmosphere below. Believed to consist of methane ice crystals, the clouds are 50 to 200 kilometers (31 to 124 miles) wide and extend for thousands of kilometres.

later, the spacecraft swung silently past Triton, a bright and frigid moon where geysers spew a mixture of gas and dust into a thin atmosphere of nitrogen.

'The image that came to mind as we stood and watched the stunning images of Neptune and Triton arrive in late August, was of standing on the prow with Columbus or Magellan or Drake as they first hove in sight of new lands here on Earth,' recounted George Alexander, JPL's manager of public affairs at the time.

Four and a half billion kilometres (2.8 billion miles) from Earth, this was Voyager 2's final encounter before drifting off into deep space. For those witnessing the flyby images on TV screens at JPL and elsewhere, it was a breathtaking experience. Neptune is the fourth biggest planet in our solar system, with a diameter 3.9 times that of Earth, yet it is so far away that it is invisible to the naked eye. Until Voyager 2 began its approach, it had been seen only by powerful telescopes, as a dim and fuzzy disc. Some scientists expected that Neptune's upper atmosphere would be indistinct, like that of Uranus. To everyone's delight, they were wrong.

Before and during its short encounter with Neptune and Triton, Voyager 2 discovered five new moons and verified the presence of a sixth that had previously been detected from Earth (bringing the total to eight). It found four rings, and like Uranus a tilted magnetic field. Operating with camera shutters open from one to several seconds to catch details of the dimly-lit planet, it determined that Neptune is cloaked with hydrogen, dosed with traces of helium and methane, and splashed with minute amounts of acetylene.

Voyager's visit to Neptune and Triton brought a magnificent close to a decade of unprecedented exploration. This odyssey by a pair of frail spacecraft brought humankind a wondrous treasure-chest of entirely new information about our own planet's four larger planetary neighbours. Its interplanetary journey ended, Voyager 2 dipped below the ecliptic, leaving the slim crescents of Neptune and Triton in its wake. As it headed out into interstellar space, beyond the planets, Edward C. Stone of Caltech, Voyager's Project Scientist, appropriately dubbed the Neptune rendezvous 'the final movement in the Voyager symphony.'

In June, 1994, Hubble's Wiffpick camera captured high resolution images of Neptune, described in a 1995 announcement from the Space Telescope Science Institute. Earlier terrestrial observations led to a model of a planet containing mostly hydrogen and helium, coloured blue by traces of methane, with a few clouds. Neptune has somewhat more methane in its atmosphere than Uranus, hence its stronger colour. The planet is also thought to have a solid core, similar to that of Uranus. Voyager 2 added to the model strong winds, approaching 1,600 kilometres per hour (1,000 miles per hour) at the equator, bright clouds, two dark spots that were apparently storm systems – the larger dubbed the Great Dark Spot – and a nearby bright area called the Bright Companion. In the new Hubble images, both of the dark spots had vanished without trace. There remained a bright cloud on the equator, two broken bright bands on the southern hemisphere, and one broken band in the northern hemisphere. It was as though the whole planet had been becalmed.

In April 1995, Hubble researchers found a new dark spot in the northern hemisphere of Neptune, as if the Voyager 2 feature had reappeared, and theorized that it might be a hole in the methane cloud tops. Then a series of near-infrared observations with the CFHT in Hawaii showed a complex atmosphere with layers of high-altitude clouds and low-altitude haze. Obviously this was a much more dynamic planet than was first suspected.

'That's what makes Neptune a special and interesting place to study,' said Hammel in a note to friends and colleagues. 'Its clouds change dramatically, very quickly.'

Of the planet's eight moons, at least two exhibit some rather odd behaviour. Nereid, which appears to be only about 340 kilometres (210 miles) in diameter, has a very irregular orbit, passing within 1.4 million kilometres (900,000 miles) of Neptune at its closest point and receding as far as 9.7 million kilometres (6 million miles) at its farthest point.

But Triton, though slightly smaller than Earth's moon at 2,705 kilometres (1,681 miles) in diameter, is the star of the Neptunian system, and was one of the main destinations of the Voyager explorations (see box on p.244).

Two of Neptune's rings are mysterious in that some regions contain more material than others. At one time it seemed that the thicker parts were disconnected, forming three main clumps or arcs named (in deference to LeVerrier) for the French revolutionary slogan Liberté, Egalité, and Fraternité. Inspection of the rings backlit by the Sun showed that they were in fact joined by arcs of thinner material. No one has adequately explained why the rings are not of similar density all the way around.

Neptune's visible cloud deck has a temperature of about -214°C (-350°F). Spacecraft can make these measurements with radiometers – instruments that detect temperatures from the infrared radiations emitted from the planets. Even better temperature measurements can be made with spectrometers, by analysing the wavelengths of light emitted from planetary atmospheres. But the best thermometers of all are radio transmissions, which appear to shift

Triton

Reporters covering Voyager 2's Triton fly-by were as astonished as anyone else: there was no need to 'popularize' their prose, for this moon said it all. It was, as Scientific American's June Kinoshita wrote, a world unlike any other, with a texture that was described variously as being like tripe, like cantaloupe, and even like cellulite.

Earthbound observers had been astounded by a tantalizing slice of complex geology – a view of rugged, jagged terrain coming up to the shores of an icy sea. It was a mind-bending experience to see mountains and valleys in a place 4.5 billion kilometres (2.8 billion miles) from Earth.

It's a fascinating place in many ways. Triton moves the 'wrong way' around its parent planet, orbiting clockwise while Neptune, as viewed from above its north pole, turns counterclockwise on its axis. Its nitrogen atmosphere – what little there is of it – is transparent. The first computer-enhanced photographs of Triton's surface, mottled in gorgeous hues of pink and blue, were spellbinding. It was something from science fiction – a sea of frozen natural gas (methane) covered in strange pockmarks, especially at the craggy south polar regions. Water ice probably lies under the nitrogen, and beneath that there is probably a molten core. Dark-coloured geysers shoot to nearly 8 kilometres (5 miles) above the surface, leaving a trail of smog. While Voyager's images were not sharp, they were made from two different angles, so that the narrowness and height of this vertical jet can be easily confirmed, as can the height of the smog trail.

Where do the geyser plumes come from? How are they generated? Nobody knows, but Laurence Soderblom of the U.S. Geophysical Survey in Flagstaff, Arizona, suggested that some natural solar collector may exist on the moon's surface. Such a collector, composed perhaps of dark, energy-absorbent material, might serve to heat up sub-surface reservoirs of material to the point where they erupt. Duncan Steel of the University of Salford, England, suggests that the plumes are actually ice volcanoes, generating syrupy lava flows of ammonia and water mixed together (see p.252).

In 1966 a graduate student at California Institute of Technology, Thomas McCord, theorized that immense tides between Neptune and Triton exert gravitational drag on the large satellite. At some time in the past, these tides may have caused Triton to drop from a more distant orbit, throwing Nereid into its present elliptical trajectory with an overwhelming jolt of gravitational force. It is still believed that Triton assumed its retrograde 'backwards' orbit

when it was captured by Neptune's gravity, and that it kicked other satellites out of the way as its orbit circularized.

Since Triton's orbit should continue to decay owing to tidal drag, McCord reasoned, it is likely to collide with Neptune's surface sometime within the next ten million years. Alternatively, it might break up into millions of tiny pieces, forming rings of debris similar to those around Saturn.

A Voyager 2 mosaic image of Triton. The south polar region shows numerous elongated dark streaks or plumes up to 60 kilometres (40 miles) long. They are thought to be the wind-blown deposits of material that erupted from geysers or ice volcanoes.

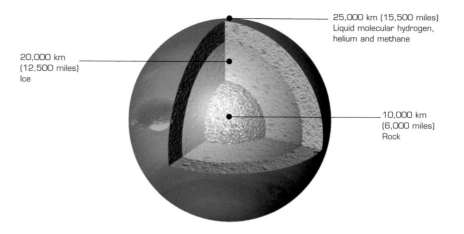

25,000 km (15,500 miles)
Liquid molecular hydrogen, helium and methane

20,000 km
(12,500 miles)
Ice

10,000 km
(6,000 miles)
Rock

A possible model for Neptune's interior. Recent research suggests that it is possible that Neptune may have a liquid core.

wavelength as they pass through planetary atmospheres just before, and just after, the spacecraft passes behind the planet. This phenomenon is caused by refraction, which itself changes with temperature.

Scientists know the masses of the outer planets quite precisely from their reaction to the laws of celestial mechanics. The mean density, however, is calculated from mass, radius, and polar flattening. Unfortunately, in pre-Voyager days the outer planets were so far away that it was impossible to make out a sharp edge to their telescopic images, and our estimates of radius were therefore greatly subject to error. We knew this meant the reference books' quoted planetary densities were probably inaccurate–and more importantly that we lacked one of the vital elements necessary to build 'models' of planetary atmospheres.

In 1965, for example, the mass of Neptune was figured at 17.2 times that of Earth, its radius at 22,600 kilometres (14,000 miles), and its polar flattening (oblateness), 0.02 per cent. This gave us a mean density around 2.2 times that of water (in other words, 2.2 grams per cubic centimetre).

But in 1968, scientists were able to watch as a star was gradually obscured (occulted) by Neptune, and to note its position as it slipped behind the visible edge of the planet's disc. When they measured a line drawn from this spot to the centre of the disc, they learned that the radius was greater than they had thought – and that therefore Neptune's mean density must be less than previously estimated. Today we know the radius is actually 24,760 kilometres (15,380 miles), and the mean density has been edged down to 1.64 g/cm^3 (97 lb per cubic foot).

This type of measurement triggers much activity among planetologists. If the mean density is lowered, estimates of the abundance of hydrogen and helium compared with the heavier elements may be raised.

Thanks to Voyager, we know a great deal more about the planet as seen from outside its atmosphere. We know, for example, that the wind in the planet's cloud tops can move at close to the speed of sound, and that some bands of wind move from east to west, and others from west to east. But what lies beneath that cloak of cloud? And where does the planet get its energy?

At the outermost limits of the solar system, completing its orbit once every 165 years, Neptune has an as-yet-unidentified internal heat source that generates its strange weather system. In fact it radiates 2.6 times more energy than it receives from the Sun – still not much, for the planet receives only 1/900 the amount of solar energy received by Earth. But it's a mystery. And from this mystery comes a possible answer, and with it a model of the planet's interior.

In 1999, researchers at University of California at Berkeley decided they would attempt to replicate conditions about a third of the way down into Neptune's mass. They squeezed a sample of methane – molecules that consist of four hydrogen atoms around a carbon atom – in a high-pressure diamond anvil press. Then they zapped it with a laser beam that had been carefully tailored to replicate the pressure and temperature present at that depth.

What happened next made headlines around the world. The methane simply fell apart, and the carbon atoms combined under high pressure. The result: diamond dust. It's theorized that friction from a rain of diamonds from the outer regions of Neptune towards its core could account for a good part of the energy that the planet radiates.

PLUTO:
Most Mysterious of All

'Ninth planet discovered on edge of solar system,' announced the front page of the *New York Times* for March 14, 1930. 'First found in 86 years... lies far beyond Neptune.'

Below seven decks of headlines, an Associated Press dispatch filed the previous day from Flagstaff, Arizona, described the discovery in florid terms:

'In the little cluster of orbs which scampers across the sidereal abyss under the name of the solar system there are, let it be known, nine instead of a mere eight, worlds. The presence of a ninth planet in the retinue of the Sun, long suspected, was definitely announced here today.'

At the Lowell Observatory at Flagstaff, the news was already almost two months old. For many weeks, the observatory's staff of astronomers had kept the discovery quiet while they checked and rechecked the position of this mysterious Planet X. Then they made the news public in an announcement to Harvard Observatory. The date was March 13, 1930, the 149th anniversary of the discovery of Uranus and the birthdate of Percival Lowell (1855–1916), who had predicted the existence of the new planet many years earlier, in 1905.

Planet X – later named Pluto – was discovered by a young scientist named Clyde W. Tombaugh (1906–97), who had developed an interest in astronomy while working on his father's farm in Illinois. His drawings of Jupiter and Mars, made with the aid of a telescope he built himself, led to an appointment on the staff of Lowell Observatory. There, he was promptly assigned to the search for undiscovered planets beyond Neptune.

Tombaugh then began a painstaking search of sky photographs with the help of a 'blink comparator' – a device through which the viewer looks at two photos of the same part of the sky, taken at different times. When the views are alternated rapidly, any image that has moved with respect to its neighbours seems to jump back and forth. Thus planets, asteroids, and comets are more readily identifiable against the background of stars.

On the afternoon of February 18, 1930, Tombaugh recalled, 'I suddenly spied a 15th-magnitude object [that is, something 4,000 times fainter than the faintest naked-eye star] popping in and out of the background. Just three millimetres [0.1 in] away another 15th-magnitude image was doing the same, but appearing alternately with respect to the other, as first one plane and then the second was visible through the eyepiece. 'That's it!' I exclaimed to myself...'

The story started in the early years of the twentieth century, with Percival Lowell, the mathematician-astronomer who proposed that a doomed civilization had built canals on Mars. Lowell suspected that another planet must lie beyond Neptune because irregularities had been reported in that planet's orbit. His reasoning was that, if Neptune's presence had been indicated by irregularities in the orbit of Uranus, then perhaps history would repeat itself with Planet X. Other scientists felt that the indications of Neptune's irregularities were no more than 'noise in the system,' but not Lowell. His search for the ninth planet began in earnest in 1905, at his personal mountain observatory near Flagstaff, Arizona.

Soon a Harvard astronomer, William H. Pickering (1858–1938), would also be looking for a trans-Neptunian planet that he called Planet O, one of a string of hypothetical planets up to and including the letter U.

Above: Pluto is the only planet not yet visited by a spacecraft. But some features of its surface have been unveiled by the Hubble Space Telescope's Faint Object Camera (FOC). These features are probably caused by the complex distribution of frosts that migrate across Pluto's surface with its orbital and seasonal cycles.

Left: This is the clearest view yet of Pluto and its moon, Charon, as revealed by the Hubble Space Telescope. Hubble's ability to distinguish Pluto at a distance of 4.4 billion kilometres (2.6 billion miles) is equivalent to seeing a tennis ball at a distance of 64 kilometres (40 miles). Prior to Charon's discovery in 1978, it was thought that Pluto was much larger because our images blurred the two bodies together into one 'planet'.

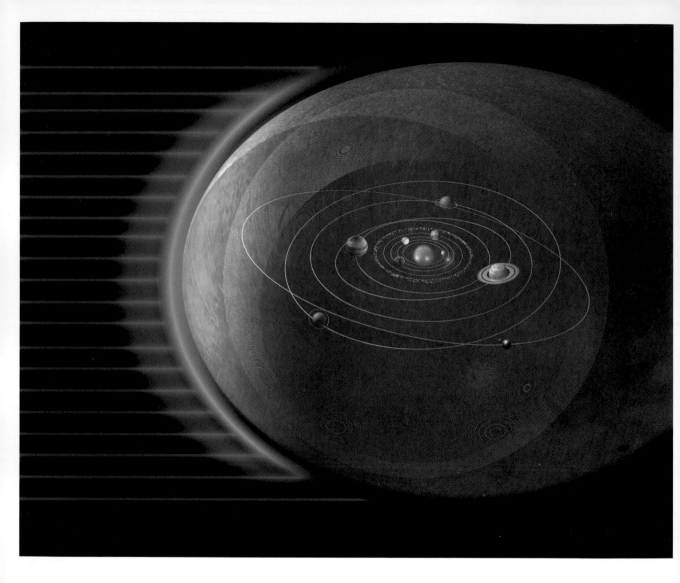

The heliopause marks the edge of our solar system and the beginning of interstellar space. At 1.6 million kph (1 million mph), the solar wind hurtles into space, creating an expanding bubble of plasma (the heliosphere) around the Sun. Where the pressure of the outer edge of the heliosphere equalizes with the interstellar medium, lies the heliopause. We believe the heliopause is between 90 and 120 A.U. from the Sun, which Voyager 1 should verify in the next few years.

Lowell died in 1916, but his work went on under Vesto Slipher (1875–1969), who hired a 'young man from Kansas,' Tombaugh, to continue the search.

Ironically, neither Lowell, Slipher, nor Pickering would be credited with finding this peculiar little planet. Tombaugh, exercising incredible dedication and patience, eventually made the discovery.

Pluto is named for the Roman god of the underworld. This is appropriate, for Pluto remains the most mysterious of the planets. One fifth the diameter of Earth, but with just one fifth of one per cent of Earth's mass, it is an oddball indeed. Or maybe, more accurately, it is two oddballs. Pluto's single moon, Charon, was detected in 1978 by James W. Christy of the U.S. Naval Observatory at Flagstaff, after he noticed an occasional slight distortion in the image of Pluto. Charon is by far the largest moon in the solar system compared to its parent planet, and some astronomers consider the pair a 'double planet'.

Quite likely Charon was formed in the same way as Earth's moon. Chunks of matter torn from Pluto by, say, an early asteroid crash, might have been captured by the icy planet's gravitational field, then gradually come together to form a single, spherical moon.

In 1936, the British astronomer Raymond A. Lyttleton (1911–95) suggested that Pluto was once a satellite of Neptune, but was flung out of orbit during a close encounter with Triton that also reversed that moon's orbit. This was at first an attractive theory, but was discarded after closer scrutiny.

Current theory holds that Pluto originally had a circular orbit, but gradually became eccentric as it was forced into a 3:2 resonant orbit by Neptune's gravity. For every three times that Neptune circles the Sun, Pluto makes the circuit exactly two times. This saves the smaller planet from moving too close to its larger neighbour and being captured. 'Pluto,' wrote planetary physicist Renu Malhotra, 'is engaged in an elegant cosmic dance with Neptune, dodging collisions with the gas giant over the entire age of the solar system.'

Is Pluto really a major planet that deserves equal billing with its eight brethren? It is too small and lightweight to be a terrestrial planet like Earth, Mars, Venus, and Mercury. It is clearly not a gas giant. It, and its moon, are icy like the population of the Kuiper Belt. So either Pluto is the 'King of the Kuiper Belt' or it belongs to an in-between category.

If planets are classified simply as spherical bodies that orbit a star (a definition that excludes irregularly shaped asteroids and comets), then Pluto certainly qualifies. The naysayers gained a big supporter in early 2001, when the Hayden Planetarium in New York declared Pluto a non-planet, mostly because it did not qualify as either a terrestrial or a Jovian planet. To them it was more like a refugee from the Kuiper Belt. In fact, many astronomers consider Pluto and its moon, as well as Triton, to be Kuiper Belt Objects (KBOs), and even surmise that similar icy planets may lie farther out in the belt.

Though no spacecraft has yet come close to Pluto, we have learned a lot from telescopic observations – from sites on Earth as well as from Hubble. We were somewhat surprised to learn that, rather than being an inert snowball, Pluto has a dynamic atmosphere. Apparently credit is due to its relatively oval-shaped, highly elliptical orbit. It takes Pluto 248 Earth years to complete that orbit. Pluto's most recent midsummer's day was in late 1989. Its coldest midwinter day was in 1865; the next will be in 2113. A fresh surface of methane

Dr Duncan Steel
Space Researcher
University of Salford

Ices in the Solar System

The word 'ices' can mean several different things to an astronomer. Apart from water, several other compounds come under that heading. One will be familiar to all: the solid carbon dioxide that has many commercial uses, and is often called 'dry ice.'

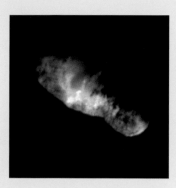

French astronomer Alphonse Borrelly discovered this comet in 1904. Since then its returns have been viewed several times – because it laps the Sun every seven years – but no ground-based photographs could match what was seen by the Deep Space 1 satellite as it passed within 2,250 kilometres (1,400 miles) of it on September 22nd, 2001. The nucleus was found to be about eight kilometres (five miles) long, half of that wide, and shaped like a bottle. Erupting from this solid core were jets of dust and vapour, up to 100 kilometres (60 miles) long. These then dispersed to make the vast cloud surrounding and obscuring the nucleus when studied from afar, and eventually formed the tail of the comet. Although comets appear bright in the night sky, that is due to the scattering of sunlight by their cloud and tail. Their solid nuclei are intrinsically dark. This is especially true for Borrelly, which reflects only a few percent of the sunlight striking it, making it blacker than asphalt. If it seems surprising that a lump largely made of ice could be so dark, try mixing a cupful of soot into a bucket of pristine white snow: the result is always murky.

Various ices are important in planetary science. Imagine a comet entering the solar system on an elongated orbit. Many such 'new' comets announce their existence when they radically brighten on crossing a solar distance of three astronomical units.

Why? The reason is simple: comets are largely composed of water ice, and the physical properties of water dictate that it remains solid when outside 3 A.U., where the sunlight is too weak to cause its sublimation (direct change from solid to vapour). Once the threshold is crossed water starts to vaporize. This forms a coma (the huge cloud around the solid nucleus, reflecting a great deal of sunlight), and also a fledgling tail. It is with this abrupt brightening that comets may be easily discovered.

But other comets are observed – albeit as faint objects – in the deeper reaches of the solar system. There, they may sport tenuous but definite comas. How come?

The answer lies with the other ices, which sublimate at lower temperatures than water. Using their physical parameters, measured in a laboratory, we can calculate the heliocentric distances at which different ices might vaporize, and cross-check that against the distances at which comets actually produce vapour clouds. Carbon dioxide (CO_2), for example, has a critical distance of 13 A.U. from the Sun, whereas its sibling carbon monoxide (CO) is less stable, vaporizing much further out. Similarly ammonia (NH_3) has a critical distance of 11 A.U.

We do see comets being active (that is, forming weak comas) at these sorts of distances, and spectroscopic observations have confirmed the presence of these molecules in their comas.

The above ices are all inorganics. We know that comets also contain large quantities of another type of ice: organic compounds. Methane (CH_4) sublimates at 84 A.U., ethane (C_2H_6) at 24 A.U., and the familiar gas fuel propane (C_3H_8) at 14 A.U. Proceeding up this sequence of molecules (the alkanes), it is not until you reach octadecane ($C_{18}H_{38}$) that you find a compound that is stable at the Earth (i.e., at 1 A.U.).

Many common organics might also be mentioned, such as ethyl alcohol (C_2H_5OH), acetone (C_2H_6CO), and benzene (C_6H_6): all are unstable within 3 to 5 A.U. With Chris McKay of NASA-Ames Research Center I have investigated the stability of over a hundred organic compounds in comets, with a view to explaining how such chemicals may have been supplied to the primordial Earth, providing the basic building blocks of life.

Comets contain lots of ices, then, but they are also found elsewhere. The polar caps of Mars are largely carbon dioxide. Jupiter's moon Europa has an icy crust. Two other Galilean moons, Callisto and Ganymede, also have major complements of ice. Saturn's Titan has a thick atmosphere of ethane and methane, and many of its smaller satellites are icy.

The ices really come into their own in the far reaches of the planetary system, though. So far from the Sun, the ambient temperature drops below -200°C (-330°F), and molecules we think of as being gases (like ammonia, carbon dioxide, and ethane) are solids.

Neptune's moon Triton is a good example. Triton is bright (it was found within months of Neptune itself being discovered in 1846) because of highly reflective frost coating its surface. This must come and go. The scarcity of impact craters also bears witness to its frequent surface renewal.

The only other moon with this property is Io, which has a surface constantly replenished through volcanic action, sulphur and other compounds being spewed out from its hot interior. Triton also has volcanoes – but they're ice volcanoes. When Voyager 2 whizzed past Triton in 1989, it imaged plumes up to eight kilometres (five miles) high.

Laboratory studies have indicated how lava can be formed at such low temperatures, but this lava is not the familiar molten rock – it's made of ices. What is known as a 'eutectic melt' mix of ammonia and water is a fluid at -100°C (-148°F), flowing in much the same viscous manner as honey or molasses. Triton had been expected to be a dead, frozen world; it may be freezing, but geologically-speaking it is very much alive, due to the peculiarities of ices.

We've yet to get a close-up view of Pluto and its partner, Charon. There have been plans advanced in the U.S. for a Pluto probe, but it has been very much a case of on-again, off-again due to NASA budgetary constraints. Such a mission is time-critical, again for reasons connected with the behaviour of ices.

Between 1979 and 1999 Pluto was not the outermost planet, being slightly nearer the Sun than Neptune. The significant eccentricity of Pluto's 248-year orbit means that its heliocentric distance varies between 29.7 and 49.4 A.U. When closest in, the ices of which it is largely composed form a significant atmosphere. But as Pluto recedes it cools, and that atmosphere freezes out, which is happening now. If we miss the chance to study Pluto's atmosphere within the next ten to fifteen years, we'll have another two centuries to wait for our next opportunity.

A huge number of minor planets (or asteroids) orbit near Pluto in a region called the Kuiper Belt. The first member of this trans-Neptunian band was spotted in 1992, and since then hundreds have been found, ranging upwards in size from 100 kilometres (60 miles). Some are bigger than Ceres, the largest asteroid in the main belt between Mars and Jupiter.

It is a moot point, though, whether these trans-Neptunian objects (TNOs) should be classified as asteroids. They are largely icy bodies, and so in essence they are giant comets.

Apart from the large TNOs observable from Earth, there must be greater numbers of small ones. This reservoir is regarded as the source of most short-period comets (like Halley's and Encke's comets). The Kuiper Belt is a flattened disk, producing the low orbital tilts of most short-period comets.

Long-period comets, arriving on near-parabolic paths, are a different matter entirely. These fall into the inner solar system, with random orientations, from the Oort cloud, a spherical distribution 10,000 to 100,000 A.U. away. Their ejection stems from perturbations induced by passing stars and interstellar clouds, and by the galaxy's tidal force. But how did those comets originally arrive in the Oort cloud?

The physical behaviour of ices once more provides a clue. Various lines of evidence, including the nuclear temperatures of long-period comets, suggest they were formed in the Uranus-Neptune region. Those planets then threw the comets outward into the reservoir where they have resided the past 4.5 billion years. Paradoxically, then, the Oort cloud comets seem to have solidified closer in than did the Kuiper Belt objects.

There are many questions we need to answer, but one thing is clear: When it comes to understanding the solar system's formation, ices can tell us a lot.

ice freezes onto its surface as temperatures cool during that 124-year journey away from the Sun. 'Dirty' trace elements mark the white surface, but evaporate into the atmosphere during Plutonian summer, when temperatures rise to -210°C (-350°F). It's likely that high winds are generated in the process.

Pluto is believed to have a rocky and/or icy interior, surrounded by a thick mantle of water ices, topped with surface ices composed of nitrogen, methane, and carbon monoxide. A 1988 occultation experiment mapped its atmospheric temperatures and detected a haze layer near the surface. Surface maps showed a pinkish tint around the planet's equator – perhaps material produced from the frosts of nitrogen, methane, and carbon monoxide that seem to lie nearby. Later, blue and yellowish-red regions were found within the coloured band, clues perhaps to the method by which organic reactions occur on the ice.

Inspection of Charon's disc with Hubble's NICMOS showed again that, unlike its larger companion, this is a body coated with pristine ice.

Pluto will be the last major body in our Solar System to be seen close-up by a spacecraft. One of the Voyager spacecraft could have been sent there directly from Saturn, but mission planners thought the price was too great. Pluto enthusiasts then argued for a special flight to the most distant planet, pointing out that a 2004 or 2006 launch could take advantage of a large-scale gravity assist from Jupiter. After that, the opportunity would lose its lustre as the planet moved outward from the Sun on an outward leg of its long elliptical orbit. It would get so cold that its thin atmosphere of nitrogen and methane would likely freeze and settle on the surface; in fact, the atmosphere could disappear in this fashion by the year 2020.

One of the most exciting challenges for a future spacecraft mission will be to scan Pluto's surface with imaging instruments. Unlike the other outer planets, it has a very thin atmosphere, so that we may be able to see a solid surface instead of cloud tops.

Beyond Pluto

Beyond the orbits of Neptune and Pluto, beginning at about six billion kilometres (3.75 billion miles) from the Sun, lies the disc-shaped Kuiper Belt, populated with hundreds of millions of icy remnants left over from the dawn of the solar system.

The belt's existence was first proposed in the 1940s and early 1950s by Gerard Kuiper and the British astronomer Kenneth Edgeworth (hence it is also called the Edgeworth-Kuiper Belt). David Jewett of the University of Hawaii and

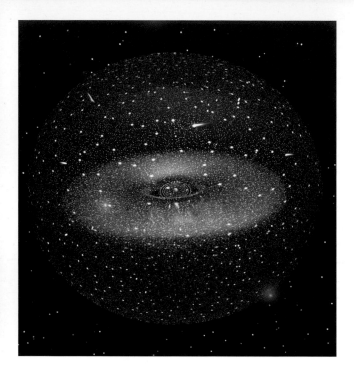

Comets are the most abundant objects in the solar system and come from both the disk-shaped Kuiper belt and the spherical Oort cloud. The Kuiper belt lies beyond the orbits of Neptune and Pluto between 30 and 100 A.U., and is the source of short period comets. The Oort cloud lies at the very edge of the Sun's gravitational influence – so far away that we can only infer its existence from indirect evidence. Long period comets such as Hyakutake or Hale-Bopp are thought to have originated in the Oort cloud.

Jane Luu of UC Berkeley discovered the first Kuiper Belt Object (KBO) with the University of Hawaii's 2.2-metre (88-inch) telescope in 1992. Planetary scientist Alan Stern of Southwest Research Institute typified the difficulty of finding individual KBOs as looking for 'something the size of a mountain, draped in black velvet, located four billion miles away.'

At this writing, about 400 KBOs had been tracked and identified. They fall into three categories. Those with a similar orbit to Pluto, in 3:2 resonance with Neptune, are called plutinos. 'Classical' KBOs are confined to the disc structure. And then there are scattered disc objects, which have highly eccentric orbits.

In 2001, researchers at the European Southern Observatory found a KBO that could be 1,200 kilometres (750 miles) or more in diameter, just beyond Pluto and Charon's orbit. If this estimate is correct, then Ixion (or 2001 KX76) is larger than both Charon and Ceres, the largest body in the asteroid belt.

Beyond the Kuiper Belt, some 150 billion kilometres (92 billion miles) from the Sun, lies a great cloud of comets, the Oort Cloud, proposed in 1950 by the Dutch astronomer Jan Oort. While theoretically the Oort Cloud may make up a significant fraction of the mass of the solar system, with as many as a trillion comets, we have no direct evidence to decide the matter.

In some respects the early solar system was like a giant concrete mixer, filled with debris from the original nebula. Collisions and near misses were frequent. Calculations indicate that the Oort cloud comets are survivors from this protoplanetary mixer, flung into their remote orbits by close encounters with the gravity of the giant planets. Stern and Paul Weissman of JPL estimate that the Oort Cloud contains mass equal to a few Earths, or even less. Previous estimates were 10–40 Earth masses.

Every once in a while, a Kuiper Belt body will collide with another, or be disturbed by the gravitational influence of a giant planet, so that it is nudged well inside the orbit of Neptune and becomes a 'Centaur'. Eventually these icy objects are either spun out of the solar system or sent into an elliptical orbit around the Sun. In the latter case we may be presented with another Halley or another Comet Kohoutek. The Centaur Chiron, at 170 kilometres (105 miles) diameter, is 20 times larger than Halley, and would turn into an extraordinary comet if it ever approached the Sun.

The Kuiper Belt is of special interest to astronomers because it has been preserved from external influences since the beginning, its objects are pristine remnants of the original solar nebula. As Stern told a meeting of the American Astronomical Society in 2000, 'This region is planetary science's equivalent of an archaeological dig into the history of the ancient outer solar system.'

Sometime in this century, a lone spacecraft will leave the far reaches of the solar system and head towards the Kuiper Belt after sending the first detailed reports of Pluto and Charon home to Earth. Until then, we have cause to reflect with anticipation upon the words of James Christy, discoverer of Charon:

'In Greek myth, souls were sometimes kept waiting many years on the shores of the river Styx until Charon took them across into the afterlife, the domain of Pluto. Today, Pluto and Charon still stand guard at the boundary of the Solar System, challenging us to discover, daring us to draw new mysteries out of the darkness.'

And so we come to the end of the planetary system as we know it. We have learned that, on a cosmic scale, it is quite small, and that we, as Carl Sagan put it, are 'starstuff pondering the stars; organized assemblages of ten billion billion billion atoms considering the evolution of atoms; tracing the long journey by which, here at least, consciousness arose.'

There is philosophy, and hope, in this great enterprise. What a tribute to the human mind it is! We are replete with ambitions and egotisms, and with our love of the temporal – and yet we project our consciousness to the moons of Jupiter and light-years beyond, to the edges of the universe. This fascination, which has endured in some form since our ancestors first contemplated the stars, has nothing to do with the material. It exists because we already possess the greatest treasure of all, wonder. For such a species, there is hope indeed.

'The universe has been extended outwardly from our own galaxy, the star group of our Milky Way, to enormously greater reaches of space and inwardly from the periphery of the atom to entities a million times smaller. The wonder is that the human mind can conceive at all of these realms so remote from the environment in which it has had its slow development over a period of a thousand centuries.'

Victor Guillemin

Freeman Dyson
Institute of Advanced Studies

Visions of the Future

When I published my speculations about the future of the universe in the Reviews of Modern Physics, the main argument was a chain of mathematical calculations beginnings with equation 1 and ending with equation 137. I cannot explain here everything that went into the equations. Instead, I will summarize briefly the results that came out in the end.

I was attempting to find answers to three questions within the framework of an open and indefinitely expanding universe.

(1) Does the universe freeze into a state of permanent physical quiescence as it expands and cools?

(2) Is it possible for life and intelligence to survive forever?

(3) Is it possible to maintain communication and transmit information across the constantly expanding distances between galaxies?

Tentatively, I answer these three questions with a no, a yes, and a maybe. No, the laws of physics do not predict any final quiescence but show us things continuing to happen, physical processes continuing to operate, as far into the future as we can imagine. Yes, life and intelligence are potentially immortal, with resources of knowledge and memory constantly growing as the temperature of the universe decreases and the reserves of free energy dwindle. And maybe, intelligent beings in different parts of the universe can keep alive forever a network of communication, exchanging their ideas and constantly increasing their circle of acquaintances. The third answer is tentative because it assumes that transmitters and receivers of information can always be built with efficiency close to the theoretical ideal.

These statements are based on rough numerical estimates which may easily be wrong by a factor of ten or a hundred. Nevertheless, they give strong support to an optimistic view of the potentialities of life. They imply that the world of physics and astronomy is inexhaustible. No matter how far we go into the future, there will always be new things happening, new information coming in, new worlds to explore, a constantly expanding domain of life, consciousness and memory. I have found a universe growing without limit in richness and complexity, a universe of life surviving forever and making itself known to its neighbours across unimaginable gulfs of time.

Here on this small planet, mind has infiltrated matter and has taken control. The infiltration of mind into the universe will not be permanently halted by any catastrophe or by any barrier that I can imagine. If our species does not choose to lead the way, others will do so, or may have already done so. Mind is patient. Mind has waited for 3 billion years on this planet before composing its first string quartet. It may have to wait for another 3 billion years before it spreads all over the galaxy. I do not expect that it will have to wait so long. But if necessary, it will wait. The universe is like a fertile soil spread out all around us, ready for the seeds of mind to sprout and grow. Ultimately, late or soon, mind will come into its heritage.

What will mind do choose to do when it informs and controls the universe? That is a question which we cannot hope to answer. When mind has expanded its physical reach and its biological organization by many powers of ten beyond the human scale, we can no more expect to understand its thoughts and dreams than a Monarch butterfly can understand ours. Mind can answer our question only as God answered Job out of the whirlwind: 'Who is this that darkeneth counsel by words without knowledge?' In contemplating the future of mind in the universe, we have exhausted the resources of our puny human science. This is the point at which science ends and theology begins.

We know very little about the potentialities and the destiny of life in the universe. In speculating about these matters we follow a great tradition. Letting our imagination wander among the stars, we too may hear whispers of immortality.

[Extracted from the Gifford lectures given at Aberdeen, Scotland, April–November 1985, and published in the book *Infinite in All Directions*. (New York, Harper & Row, 1988)]

Glossary

A.U. See Astronomical Unit.

Accelerometer. Device for measuring acceleration, in metres per second per second.

Adaptive optics. Method for correcting distortions caused in telescope images by atmospheric and other effects.

Alpha particle. A positively charged particle identical to the nucleus of a helium atom.

Aphelion. Apoapsis in solar orbit.

Apoapsis. The farthest point in an orbit from the body being orbited.

Apogee. Apoapsis in Earth orbit.

Apollo. U.S. project that landed astronauts on the Moon.

Approach. The mutual coming together of two objects; i.e., a spacecraft and a planet; the final phase of a planetary-encounter mission.

Arc second. An angular measurement, equal to one-sixtieth of a minute of arc, or 1/3600 of a degree.

Ascending node. The point at which an orbit crosses the ecliptic plane going north.

Asteroid belt. The very large number of asteroids that circles the Sun between the orbits of Jupiter and Mars.

Asteroid. One of the minor planets of the solar system, less than 1000 kilometres (600 miles) in diameter but larger than meteoroids.

Astronomical Unit. A unit of length equal to Earth's mean distance from the Sun; i.e., 149,654,911 kilometres (92,935,700 miles).

Astronomy. The science of celestial bodies and their structures.

Astrophysics. That part of astronomy that deals specifically with the physical properties of celestial bodies.

Atmosphere. (1) A body of gases surrounding a planet or other celestial body. (2) A unit of pressure equivalent to the pressure exerted by Earth's atmosphere at sea level – $1.07kg/cm^2$ (14.7 pounds per square inch).

Attitude. The "tilt" of a spacecraft along one or two of its axes, with reference to another body such as the Earth, Sun, or an encounter planet.

Aurora. Fluorescent atmospheric display caused when quantities of charged particles enter the atmosphere, as at Earth's poles.

Avionics. Aviation electronics.

Axis. A straight line, real or imagined, drawn between two points; i.e., the line between the poles of a planet about which the planet rotates; or the three axes of a spacecraft, by use of which the spacecraft's attitude may be precisely described.

Azimuth. The arc of the horizon between a vertical circle passing through a celestial body, and an observer's meridian.

Barycenter. The common centre of mass about which two or more bodies revolve.

Big-Bang Theory. That theory of cosmology which holds that the universe and everything in it originated in an enormous explosion from which it is still expanding.

Binary digit. A digit, either zero or one, in the binary (two-state) system of counting common to computer and telemetry systems.

Biophilic. Favourable to life, as in the search for biophilic planets in other solar systems.

BIS. Bus interface system.

Bit. See binary digit.

Blink comparator. A device for studying the skies, allowing the examination of two photos (or sets of photos) taken of the same area at different times.

BPS. Bits Per Second, same as Baud rate.

Bus. Platform for scientific experiments and other instruments.

c. The speed of light in a vacuum – 299,792 km/s 186,206 miles/sec.

Canopus. The second brightest star in Earth's sky, located in the constellation Carina.

Carrier. The main frequency of a radio signal generated by a transmitter prior to application of modulation.

C-band. A range of microwave radio frequencies in the neighbourhood of 4 to 8 GHz.

CCD. See charge coupled device.

Celestial equator. The median line of the celestial sphere, parallel with Earth's equator.

Celestial mechanics. That branch of science that deals with the motions of celestial bodies.

Celestial sphere. An imaginary sphere of infinite radius, whose centre is the centre of the Earth. (See right ascension and declination.)

Centaur. A multipurpose upper-stage rocket powered with two LOX-liquid hydrogen engines. [2] A minor planet orbiting close to Saturn and Uranus.

Chamber. That part of a rocket motor in which gases are allowed to expand before expulsion through a throat and nozzle.

Charge Coupled Device. To convert light to electronic signals.

Coast period. The continuation of forward movement after rocket cut-off.

Coma. The cloud of diffuse material surrounding the nucleus of a comet.

Combustion. A chemical process, such as oxidation, accompanied by heat and light.

Comets. Small bodies composed of ice and rock in various orbits about the Sun.

Command system. The electrical and electronic system used to initiate spacecraft functions.

Concentric. Having a common centre, as rings enclosed one within the other.

Constellation. A group of stars.

Control system. A system used to guide, restrain, or otherwise direct a spacecraft.

Co-ordinate. Any one of a set of numbers required to describe the location (or direction) of a given point in space.

Corona. The Sun's extended atmosphere.

Correction. The process of changing the direction or speed of a spacecraft so that the desired trajectory is achieved.

Cosmology. The study of the origins of the universe.

Cosmos. The known and theoretical universe.

CRT. Cathode ray tube video display device.

Data acquisition. The process of collecting and recording new information.

Decay. (I) The eventual slowing-down of an orbiting body (e.g., a satellite) so that it either falls or assumes another orbit; (2) the transformation of an unstable atom (radioisotope) into a different form (see half-life).

Declination. The angular north-south measurement system used to locate objects on the celestial sphere. (See right ascension.)

Deep Space Network (DSN). NASA's world-wide system for communicating with spacecraft in deep space.

Density. The ratio of the mass of a given portion of matter to its volume. For example, the density of water is one gram per cubic centimetre.

Descending node. The point at which an orbit crosses the ecliptic plane going south.

Doping. The process of adding atoms to a semiconductor in order to give it a predominantly positive or negative charge.

Doppler effect. The change in wavelength observed when a source of waves is moving towards or away from us. In light this shifts light from approaching objects to the blue, and light from retreating objects to the red.

Doppler signature. A graphic record of changes in wavelength of telemetry signals from a spacecraft caused by the Doppler effect.

Downlink. The spacecraft-to-Earth leg of a telemetry system.

Dyne. A unit of force equal to the force required to accelerate a 1-gram mass by 1 cm per second per second. Compare with newton.

Eccentricity. The distance between the foci of an ellipse divided by the major axis.

Ecliptic. A circle describing the apparent path of the Sun across the celestial sphere.

Electron. A negatively charged elementary particle.

Ellipse. A closed plane curve for which the sum of the distances between any point on the curve and two fixed points within it (the foci) is constant.

Encounter. The close approach, or actual physical contact, of a spacecraft with a celestial body.

Energia. Heavy-lift Russian launch vehicle capable of lifting 100 tonnes to LEO.

Ephemeris. A publication giving computed positions of celestial bodies at given times; plural, ephemerides.

Equator. An imaginary circle around a body which is everywhere the same distance from the poles, defining the boundary between the northern and southern hemispheres.

Equinox. One of the two points (vernal and autumnal) at which the Sun's path crosses the celestial equator. (See ecliptic.)

ESA. European Space Agency.

Escape velocity. The minimum velocity at which a body must travel in order to escape a gravitational field; near Earth's surface, this velocity is about 11km/s or 7 miles per second. (See gravity.)

ESO. The European Southern Observatory, an intergovernmental European organization for astronomical research, with major facilities located in Chile.

ESOC. The European Space Operations Centre, home of ESA's Mission Management and Control Centre.

eV. Electron volt, a measure of the energy of subatomic particles.

Exobiology. Extraterrestrial (but not necessarily "alien") biology.

Exoplanet. A planet existing in a solar system other than our own.

Exosphere. The outermost layer of an atmosphere.

Extrasolar. Outside our own solar system.

Extraterrestrial. Beyond Earth.

Field. A region or space characterised by the presence of some force, i.e. magnetic, electromagnetic, or gravitational.

Fission. The splitting of an atom into two or more parts.

Fluorescence. The phenomenon of emitting visible light upon absorbing radiation of a shorter (usually invisible) wavelength.

Fuel. A substance that is oxidised in a rocket motor to produce the rapid expansion of gases. (See oxidizer and propellant.)

Gain. Any increase, including an increase in the strength of a telemetry or other electronic signal.

Galaxy. A group of stars (such as the Milky Way, to which our Sun belongs) that forms an individually identifiable cluster in the cosmos.

Galilean satellites. The four large satellites of Jupiter – Io, Europa, Ganymede, and Callisto. So named because Galileo discovered them when he turned his telescope toward Jupiter.

Gamma ray. High-energy, short-wavelength radiation with great penetrating ability.

Geiger-Müller counter. An instrument for detecting and measuring radioactivity.

Gemini. U.S. two-seat manoeuvrable manned spacecraft, first crewed flight March 1965.

Geostationary. A geosynchronous orbit in which the spacecraft is constrained to a constant longitude and zero latitude.

Geosynchronous. A direct, circular, low inclination orbit about the Earth having a period of 23 hours 56 minutes 4 seconds.

GHz. Gigahertz (1 billion cycles per second).

Gimbal. A type of mounting that allows an object to turn freely about any axis perpendicular to the centreline of the object.

GMT. Greenwich Mean Time, similar to Universal Time, but not updated with leap seconds.

Gravitational waves. Distortions of the space-time medium predicted by Einstein's general relativity theory.

Gravity. The effect of gravitation; i.e., the mutual attraction of any two bodies, determined by their distance and respective masses.

Gravity assist. The use of planetary gravitational fields to change the direction and speed of a ballistic trajectory. (See trajectory, ballistic.)

Great circle. An imaginary circle on the surface of a sphere whose centre is at the centre of the sphere.

Guidance. The result of commands sent to a spacecraft that move it in a desired direction.

Half-life. The time in which half the atoms in a radioactive substance disintegrate or decay to another form.

Hardware. The material parts of a spacecraft, exclusive of expendable gases or fuel.

Heat pipe. A device used to transfer thermal energy from one place to another with high efficiency.

Heat shield. Any device that protects something from heat or cold.

Heat sink. A device designed especially for the absorption of heat.

Heliocentric. Sun-related or Sun-centred.

Heliopause. The boundary theorised to be roughly circular or teardrop-shaped, marking the edge of the Sun's influence, perhaps 100 A.U. from the Sun. The point at which the solar wind is stopped or reversed.

Helioseismology. A technique for mapping solar temperature, composition, and internal movements from sound waves emitted by the Sun.

Heliosphere. The space within the boundary of the heliopause, containing the Sun and solar system. That part of the solar system that is dominated by the solar wind.

Hertz. A measure of frequency, indicating the number of cycles per second.

HGA. High-Gain Antenna onboard a spacecraft.

Horizon. The line marking the apparent junction of Earth and sky, or the point or distance, or time, beyond which something cannot be perceived.

HST. Hubble Space Telescope, managed co-operatively by NASA and the European Space Agency (ESA).

Hydrocarbon. An organic compound containing only carbon and hydrogen, such as acetylene, methane, and ethane.

Hyperbola. A curve used in spacecraft trajectories which approaches a straight line as its distance from the central body increases.

Hypergolic. Fuels that ignite spontaneously when exposed to each other.

Hz. See Hertz.

IR See Infrared.

ICBM. Intercontinental ballistic missile.

Ice dwarf. Small objects that circled the Sun during the early part of its development, now believed to exist on the outer edge of the Solar System, where some will evolve into comet nuclei.

Imaging instrument. Any instrument whose output can be converted into a image or map of the subject of interest.

Impulse. The change in momentum produced by a force, determined by the value of the force and the duration for which it is applied.

Inclination. The angle between one line and another or one plane and another, stated in degrees. Usually measured with respect to a flat horizon or a central body's equator.

Inertial guidance. A system contained within a spacecraft or missile that corrects deviations in velocity and trajectory.

Inferior conjunction. Alignment of Earth, Sun, and an inferior planet on the same side of the Sun.

Inferior planet. Planet which orbits the Sun within the Earth's orbit.

Infrared. Term meaning "below red" radiation. Electromagnetic radiation with wavelengths between about 1 micrometre and 1 millimetre; i.e. light in which the wavelength of the rays lies just beyond the red end of the visible spectrum.

Insertion. A point along a line of movement at which a spacecraft acquires the desired trajectory or first goes into orbit around a planet.

Ion. A charged particle consisting of an atom stripped of one or more of its electrons, or having captured one or more extra electrons.

IRTF. Infrared Telescope Facility, located on Mauna Kea, Hawaii.

Isotope. A variety of a given atom; as C-12, C-13, and C-14 are isotopes of carbon, carrying different atomic weights. (See radioisotope.)

Isotropic. Having uniform properties in all directions.

ISS. International Space Station.

Jovian planets. Jupiter-like planets; i.e. the gas giants Jupiter, Saturn, Uranus, and Neptune.

JPL. Jet Propulsion Laboratory, a NASA-funded institution operated by California Institute of Technology.

K-band. A range of microwave radio frequencies in the neighbourhood of 12 to 40 GHz.

KBO. Kuiper Belt Object, any of the countless icy bodies of the Kuiper Belt.

kHz. Kilohertz (one thousand cycles per second).

Kuiper Belt. A flattened cloud of comet nuclei, located beyond the known planets, near the ecliptic plane of the Solar System.

Lander. A spacecraft designed to land upon a planet, or one of its satellites, for the purpose of making scientific observations.

Laser. Light Amplification by Stimulated Emission of Radiation.

Latitude. Circles in parallel planes to that of the equator defining north-south measurements, also called parallels.

Launch. The action taken in lifting a space system from Earth's surface, the period of time during which this is effected.

L-band. A range of microwave radio frequencies in the neighbourhood of 1 to 2 GHz.

LEO. Low Earth Orbit; Low Equatorial Orbit.

LGA. Low-Gain Antenna onboard a spacecraft.

Life support. Systems providing for the body's metabolism in a spacecraft.

Light speed. 299,792 km per second (186,206 miles/sec); the constant c.

Light year. The distance light travels in vacuum in one year – about 9.5 trillion kilometres (5.9 trillion miles).

Light. Electromagnetic radiation in the neighbourhood of 300 to 700 nanometres wavelength.

Limb. The edge of the apparent disk of a celestial body.

Liquid oxygen. Oxygen super-cooled as low as -183°C (297°F), used as an oxidiser in liquid-fuel rockets.

Longitude. Great circles that pass through both the north and south poles, also called meridians.

LOX. See Liquid oxygen.

Magnetometer. An instrument for measuring a magnetic field.

Major axis. The maximum length across an ellipse.

Mantle. The layer surrounding the central core of a planet.

Mass concentration (mascon). Apparent concentrations of mass in a planet or moon, revealed by gravitational effects upon spacecraft orbits.

Mass. A measure of the quantity of matter in a body (see also weight).

Mean. Average.

Mean solar time. Time based on an average of the variations caused by Earth's non-circular orbit.

Memory readout. A spacecraft function in which information is transmitted from storage in a memory device such as a solid state recorder.

Mercury. First U.S. man-in-space programme, through which John Glenn became the first American to orbit the Earth, on February 20, 1962.

Meridians. Great circles that pass through both the north and south poles, also called lines of longitude; or through the poles of the heavens and the zenith of the observer.

Metabolism. The process by which a living organism carries out normal life functions.

Meteor. A meteoroid which is in the process of entering Earth's atmosphere, sometimes called a shooting star.

Meteorite. Rocky or metallic material which has fallen to Earth or to another planet from space.

Meteoroid. Small bodies in orbit about the Sun that may fall to Earth or another planet.

Metre. The international standard of linear measurement, equivalent to 39.37 inches.

MGA. Medium-Gain Antenna onboard a spacecraft.

Micrometeoroid. A very small meteoroid.

Microwave. Radio wave with a frequency in the GHz range.

Milky Way. The galaxy that includes the Sun and solar system.

MIPS. Million instructions per second.

Model, mathematical. A set of equations which describes the physical characteristics of a subject under study.

Modulation. The process of modifying a radio frequency by shifting its phase, frequency, or amplitude to carry information.

Module. A packaged assembly of related parts; or one of the building-block segments of a spacecraft.

Momentum. The mass of the object multiplied by its velocity, an indication of its tendency to keep moving.

Moon. A natural satellite.

Nadir. The direction from a spacecraft directly down toward the centre of a planet – opposite the zenith.

Nanometre (nm). One billionth of a metre (1 millionth of a millimetre).

NASA. National Aeronautics and Space Administration (U.S. Government agency).

Nautical Mile. equal to the distance spanned by one minute of arc in latitude, 1.852 km.

Navigation. Direction of a craft from one place to another, including determination of the craft's position and velocity.

Nebula, solar. In cosmology, the original cloud of gas and dust from which the Sun and solar system may have originated.

NEO. Near-Earth Object; also NEA, or Near-Earth Asteroid.

NERVA. Nuclear Engine for Rocket Vehicle Application, early nuclear rocket design programme.

Neutron. A particle that is one of the constituents of the atomic nucleus, and carries no charge.

Newton. A unit of force equal to the force required to accelerate a 1-kg mass by 1 metre per second per second.

NiCd. Nickel-cadmium, a material commonly used in long-lived batteries (Also called NiCad.)

NICMOS. Near Infrared Camera and Multi-Object Spectrometer.

Nodes. Points where an orbit crosses a plane.

Nova. A star that suddenly increases its light output, then quickly subsides. (See also supernova).

Nozzle. The bell-shaped aft portion of a rocket engine.

NASDA. National Space Development Agency of Japan.

NTR. Nuclear thermal rocket.

Nucleus. The central body of a comet.

Oblate. Having a slightly flattened shape (like the planet Saturn).

Occultation. Disappearance of a body behind another of apparently larger size.

Oort cloud. A nearly numberless spherical cloud of comets theorized to orbit the Sun out to distances of about two light years.

Orbit. Any trajectory; but usually a closed path described by a body in its revolution about another, as in the path of an artificial satellite around the earth, a planet around the Sun, or the Sun around the galactic centre.

Orbiter. A spacecraft designed to achieve and maintain a closed orbit around a planet or natural satellite.

Orrery. Device or software system for showing the relative position and motions of the planets.

Oxidizer. A substance that combines with another to produce heat and, in the case of a rocket, a gas. An oxidant.

Pallet. Movable or stationary platform on which spacecraft scientific instruments are mounted.

Parabola. A curve common to trajectories Unlike a circle or ellipse, it is not closed but extends to infinity.

Parallels. Circles in parallel planes to that of the equator, defining north-south measurements – also called lines of latitude.

Payload. That portion of a spacecraft or rocket vehicle designed to obtain the scientific or engineering results for which the vehicle is launched.

Periapsis. The point in an orbit closest to the body being orbited.

Perigee. Periapsis in Earth orbit.

Perihelion. Periapsis in solar orbit.

Period. The interval of time required for a periodic motion to complete a cycle.

Phase. (1) The angular distance between peaks or troughs of two waveforms of similar frequency. (2) The particular appearance of a body's state of illumination, such as the full phase of the moon. (3) Physical state of matter (gas, liquid, plasma, or solid).

Photometer. An instrument for measuring brightness, luminosity, or illumination, in infrared and ultraviolet wavelengths as well as in visible light.

Photovoltaic materials. Materials that convert light or other electromagnetic radiation into electrical current.

Pitch. "Up-down" movement of an object along its "nose-tail" axis.

Pixel. Picture element. One of a number of light-sensitive elements in an imaging instrument. Typically it will produce a signal that can be broken down into one of 256 or more degrees of brightness (shades of grey).

Planet, double. Two bodies, such as Pluto and Charon, that are of similar size and revolve around a common barycenter.

Planet, giant. Jupiter, Saturn, Uranus, and Neptune (also called gas giants).

Planet, inferior. Planets lying within Earth's orbit, closer to the Sun; i.e., Venus and Mercury.

Planet, Jovian. Any of the giant planets.

Planet, major. A planet as distinguished from natural satellites and asteroids.

Planet, minor. Term sometimes applied to asteroids.

Planet, outer. Planets lying beyond Mars' orbit, farther from the Sun; i.e., the giant planets plus Pluto.

Planet, superior. Planets lying outside Earth's orbit, farther from the Sun; i.e. Mars and the outer planets.

Planet, terrestrial. The Earthlike planets; i.e., Mercury, Venus, Earth, and Mars.

Planet. A large body that revolves around the Sun or another star.

Planetarium. A place in which representations of celestial images are projected onto a curved overhead screen, to reproduce the night sky.

Planetesimals. One of the countless small bodies, existing in the early solar system, that accreted into the various bodies of the modern solar system, including the terrestrial planets, the nuclei of the giant planets, asteroids, and comets.

Plasma. Electrically conductive fourth state of matter, consisting of ions and electrons.

Prebiotic. A state that may precede the appearance of life.

Probe, space. Any vehicle designed to penetrate outer space and send back information on conditions encountered.

Propellant. The mixture of fuel and oxidizer that ignites to provide thrust for a rocket; or any fluid flow that sets up a reaction and thus propels a vehicle.

Proton. A positive electrically charged particle found in the atomic nucleus.

Protoplanet. Any of the Sun's planets as it emerged in the formative period of the solar system.

Prototype. An early model, produced for developmental purposes.

Pulsar. A collapsed star that appears to emit radiation in pulses.

Quasar. A quasi-stellar object, an extremely luminous core seen in some distant, early galaxies.

RA. See right ascension.

Radian. Unit of angular measurement equal to the angle at the centre of a circle subtended by an arc equal in length to the radius. Equals about 57.296 (180/π) degrees.

Radiation. The emission and propagation of energy or matter.

Radioactivity. Decay or disintegration of an unstable atomic nucleus, accompanied by the emission of radiation.

Radioisotope. An unstable isotope of an element that decays or disintegrates spontaneously, emitting radiation.

Radioisotope thermoelectric generator. A device that employs thermoelectricity to produce power from the heat produced during the decay of a radioactive isotope.

Radiometer. An instrument that detects and measures electromagnetic energy.

Radiotelescope. A radio receiver used to study the sky through detection of extraterrestrial radio emissions.

Reaction wheel. A device that accomplishes changes in spacecraft attitude through transfer of momentum from itself to the vehicle.

Real time. Time in which events are reported as they occur.

Red dwarf. A small star, on the order of 100 times the mass of Jupiter.

Red shift. A displacement of light from a retreating object toward the red, as observed in the spectra of galaxies receding from the Earth. Used to estimate the speed and distance of galaxies and quasars.

Redundancy. Intentional placement of duplicate systems on a spacecraft for backup use if and when the original system fails.

Re-entry. Return of a spacecraft to Earth's atmosphere.

Refraction. The deflection or bending of electromagnetic waves when they pass from one kind of transparent medium into another.

Regolith. Fragmented dirt or soil materials on and near the surface of the Moon and Mars, formed by meteorite bombardment.

Retrograde. Rotation or revolution in the opposite direction to the rotation of the body being orbited.

RF. Radio Frequency.

Right ascension. The 'longitude' of a celestial object, measured in hours, minutes, and seconds along the celestial equator eastward from the vernal equinox.

Rocket. An engine in which thrust is achieved by the rearward expulsion of particles, without dependence upon atmospheric oxygen.

Rocket, nuclear. A rocket propelled by the expulsion of gases (i.e., hydrogen) superheated and expanded by a nuclear reactor.

Roll. A rotating motion about a spacecraft's 'nose-tail' axis.

RTG. Radioisotope Thermoelectric Generator.

Salyut. Soviet Russian space-station series, first of which was orbited in June 1971. Succeeded by the Mir station.

SAR. Synthetic aperture radar. Technique for using radar scans to make ultra-high-definition images from space.

Satellite. A small body that orbits a larger one. A natural or an artificial moon. Earth-orbiting spacecraft are called satellites.

While deep-space vehicles are technically satellites of the Sun or of another planet, or of the galactic centre, they are generally called spacecraft instead of satellites.

Saturn V. Three-stage launch vehicle developed for the U.S. Apollo programme, then retired. The largest rocket built, 109m (364 ft) high with the spacecraft capsule attached, capable of lifting about 140 tonnes into LEO.

S-band. A range of microwave radio frequencies in the neighbourhood of 2 to 4 GHz.

Scalar. Having magnitude but not direction; the term "kilometres per hour" is a scalar measurement.

Semiconductor. A material that under some conditions acts as an electrical insulator, under other conditions as a conductor.

Semi-major axis. Half the distance of an ellipse's maximum diameter, the distance from the centre of the ellipse to one end.

Sensor. Any instrument designed to gain information.

Shepherd moons. Moons which gravitationally confine ring particles.

Sidereal. Pertaining to the Sun or to the stars in general.

SiGe. Silicon-germanium, a popular thermocouple material.

Sirius. The brightest star in Earth's sky, in the constellation Canis Major. Its apparent brightness is due mainly to its relative closeness to Earth, rather than to its intrinsic brightness.

Software. Instructions used by computers to perform useful tasks.

Solar array. A collection of photovoltaic panels.

Solar system. The system of planets and other bodies that orbit the Sun.

Solar wind. The stream of charged particles radiated by the Sun across the solar system.

Solid state. Electronic devices such as transistors, which are made from solid materials; as opposed to vacuum tubes, which are contained within glass envelopes.

Space. The universe; or the areas located between celestial bodies or groups of bodies: i.e., interplanetary, interstellar, and intergalactic space.

Spacecraft. A manmade vehicle that goes into, or through, space.

Specific impulse (I_{sp}). A measure of rocket propellant performance equal to the thrust divided by the weight flow rate.

Spectrometer. An instrument for the analysis of specific parts of the spectrum.

Spectrum. The entire range of electromagnetic radiation, from gamma rays to long radio waves; including infrared, the visible spectrum, and ultraviolet.

Speed. A scalar quantity indicating the rate of change in position regardless of direction.

Stage. A section of a rocket vehicle that houses a rocket engine or motor.

Star. A luminous, gaseous body in space; the Sun is the nearest star.

Steady-State Theory. In cosmology, the theory that holds that new matter is being created in the universe, so that its essential structure is not changing.

STS. Space Transportation System (Space Shuttle).

STScI. Space Telescope Science Institute, operated for NASA by the Association of Universities for Research in Astronomy (AURA), under contract with the Goddard Space Flight Centre, Greenbelt, MD.

Superior conjunction. Alignment between Earth, the Sun, and a planet on the far side of the Sun.

Supernova. An explosion marking the end of a giant star's life cycle, which can temporarily outshine an entire galaxy.

Telemetry. The use of radio signals to transfer coded information between distant stations; as from a spacecraft to Earth.

Telemetry storage. A spacecraft function in which information is stored prior to transmission via telemetry.

Telescope. An instrument for detecting visible or other electromagnetic radiation emitted by, or reflected from, a distant object. Includes the conventional telescope, the radio telescope, and the radar telescope.

Termination shock. The boundary within the heliosphere where the solar wind slows from supersonic to subsonic.

Thermocouple. A device that converts thermal energy directly into electrical energy.

Thermoelectric Outer Planets Spacecraft.
Name given to JPL's early design and development work for a spacecraft mission to the outer planets.

Thermoelectricity. Power converted directly from thermal energy.

Throat. That part of a rocket motor in which gases are constricted prior to their escape through a nozzle.

Thrust. The force exerted on a rocket-propelled vehicle by its engine.

Thruster. An engine, generally small, used to manoeuvre a spacecraft.

Time warp. Hypothetically, a peculiarity in which the space-time continuum is warped in a way that makes it possible to short-cut from one part of the continuum to another.

Titan. [1] A three-stage launch vehicle including two strap-on solid rocket motors, available in several versions. [2] The largest moon of Saturn.

Tracking. The practice of following the progress of a spacecraft in flight, and relating its position to the Earth and other planets.

Trajectory, ballistic. A trajectory consisting mostly of unpowered flight, in which almost all the required energy is imparted at launch, or during infrequent propulsive burns.

Transducer. Device for changing one kind of energy into another – typically from heat, motion, or pressure, into a varying electrical voltage or vice-versa.

Trim manoeuvre. An adjustment, usually made with thrusters, to a spacecraft's ballistic trajectory.

Ultraviolet (UV). The "invisible" light located just beyond the shorter-wave end of the visible spectrum. Electromagnetic radiation in the neighbourhood of 10 to 390 nanometres wavelength.

Universal Time (U.T.). A twenty-four-hour timekeeping system based on the solar time at Greenwich, England, which lies on zero meridian (see GMT).

Universe. The entire, unlimited volume of creation.

Uplink. The Earth-to-spacecraft link in a telemetry system.

Van Allen belts. The fields of trapped particles extending outward from Earth.

Vector. A quantity that has both a magnitude and a direction, such as velocity.

Velocity. A measurement of speed in a specific direction.

Vidicon. Vacuum tube used for capturing video images.

Vostok. Russia's early cosmonaut programme, which put the first human in space, Yuri Gargarin, into Earth orbit on April 12, 1961.

Watt (W). A measure of electrical power.

Waveband. Any region of the electromagnetic spectrum that includes several or many individual wavelengths.

Wavelength. The distance a wave from a single oscillation of electromagnetic radiation propagates during the time required for one oscillation.

Weight. The force acting on an object with mass in a gravitational field.

WF/PC. Wide Field/Planetary Camera on the Hubble Space Telescope – nicknamed the Wiffpick.

Wormhole. Hypothetically, a passageway between one part of space-time to another part of space-time.

X-band. A range of microwave radio frequencies in the neighbourhood of 8 to 12 GHz.

X-ray. A type of electromagnetic radiation of extremely short wavelength.

Yaw. "Left-right" deviation in a spacecraft's attitude.

Zenith. The point on the celestial sphere directly above the observer – opposite the nadir.

Index

Adams, John Couch 237, 241
aerospace centres 53, 54
Airy, George 241
Alexander, George 130, 242
Alpha Centauri 110
alpha particles 23
American Institute of Aeronautics and
Astronautics 62-4
Ames Research Center 53, 87, 188, 253
antigravity 113
antimatter 110, 113
Antoniadi, Eugène M. 166
Apollo flights 55, 62, 107, 142, 158, 168
Arctic/Antarctic research 65-7, 188
Arecibo Observatory 43, 44, 167, 177
Ariane launchers 54, 116
Ariel 1 probe 105
Arrhenius, Svante A. 175
Asbolus 74
Asimov, Isaac 98
ASTER imaging 149
asteroid belt 37, 68, 69, 77, 101
    second belt 174
asteroids 15, 68, 69-73, 84, 225, 254
Astrium 54, 151
astrobiology 87, 192-4, 211, 221
astronomical unit (A.U.) 25
Atlas-Agena launch system 134
auroras 32-4, 52, 153, 200, 205, 233

bacterial life 211, 224-7
Barnard, Edward 213
Bartoli, Daniel 200
Beer, Wilhelm 76, 183
BepiColombo spacecraft 175, 181-2
Bessel, Friedrich W. 237, 241
Bethe, Hans 21-4
Big Bang theory 16, 17-19
binary code 123, 135
black holes 19
blueshift 16
Bode, Johann Elert 77, 230
Bondi, Hermann 16
Boyle, Charles 78
Boys, Charles C. 147
Bradbury, Ray 61
Braham, Stephen 65-7
Braun, Karl Ferdinand 147
Braun, Wernher von 55, 98
Brownlee, Donald 142-4
Bruno, Giordano 43, 137

cameras 39, 47, 49, 249
carbon dioxide 140, 175, 185-6
Carrington, Richard C. 26
Cassini, Giovanni D. 25, 183, 200, 216, 222
Cassini mission 52, 56, 85-6, 222
    design and equipment 91-5, 100-101,
    107-11, 117, 118-20
    information gathering 120-24
    mission controllers 129-30
    see also Huygens probe
Cavendish, Henry 147
celestial mechanics 43, 83
celestial sphere 76
celestial timetables 74-9
Centaur launch systems 107, 111, 117
Centaurs 74, 223, 256-7
Cerenkov radiation 22
Ceres 70, 254, 256
Chaco Canyon, NM 38-9
charge-coupled devices (CCD) 47, 100
Charon see Pluto, moon
Chicxulub impact crater 71, 155
Chinese space research 64, 113, 152
Christy, James W. 250-51, 257
Cicero 156
Clarke, Arthur C. 38, 55, 57-8, 98, 125, 148,
197, 229
Clementine spacecraft 158, 160
Cluster missions 27, 28, 156
comets 15, 68, 74, 111, 238, 252-4, 256
    fragmented 205
    Halley's 38, 74
    impacts from 70, 191
command and data systems (CDS) 94-5
computing and telemetry 120-35
confinement problems 64
Copernicus, Nicolaus 40, 43, 183
Corliss, William H. 28
corona of Sun 27
coronograph 26
cosmology 16-19
COSTAR 51
Crab Nebula 35, 38
Cygnus 19

dark energy 16
Darling, David 192-4
Darwin, Charles 227
Deep Space 1 111, 130, 252
Deep Space Network (DSN) 90, 120, 125-32
Deneb 19
Deslandres, Henri A. 26
deuterium 17, 22
Devon Island 66-7
diamond dust 247
Digital Orrery 78-9

dinosaur extinction 71, 225
distance measurement 25, 37-8, 77, 132
Doppler shifts 16, 45, 124, 177, 178-9
Dunham, Theodore 219
Dunkin, Sarah 181-2
Dyson, Freeman 62, 258-9

Earth 68, 136-61
    circumference calculation 145
    climate 40, 140-41, 151-3
    density 147
    early maps 39, 145-8
    ecosphere and magnetosphere 155-6
    how humans affect our planet 140-41
    how rare is the Earth? 142-4
    life created on 140, 224-7, 253
    make-up of planet Earth 153-6
    mass 147, 153
    moon see Moon
    plate tectonics 40, 142, 147, 154
    remote imaging 147-53
    as seen from Moon 142
    as seen from space 136-9
    solar materials in core 38
Earth-like planets, possibilities of 51, 145,
191-5
eclipses 19-20, 33, 39, 40, 145, 174
Eddington, Arthur Stanley 21, 33
Edgeworth, Kenneth 255
Einstein, Albert 33, 174
electromagnetic radiation (EMR) 23
electrons 23
Encke, Johann Franz 216
engineering flight computers (EFCs) 94
ephemerides 74-9
Eratosthenes 145
EROS Data Center, Iowa 152
Europa 57, 192, 194, 209-11, 221, 253
European Space Agency (ESA) 51, 52, 54,
64, 74, 90, 131, 151, 194, 197, see also
BepiColombo; Huygens probe; NASA/ESA
European Space Operations Centre (ESOC)
54, 120
Explorer 1 96, 148
extrasolar planets (exoplanets) 45-6

**Far Ultraviolet Spectroscopic Explorer (FUSE)
satellite** 17, 196
Faraday, Michael 124, 134
Fraunhofer, Joseph von 25, 47, 219

galaxies, ancient 17
Galileo 25, 39, 40, 43, 46
    and giant planets 39, 46, 200, 215-16
    and terrestrial planets 39, 166, 175, 183
Galileo spacecraft 52, 56, 132, 203, 205-9
    design and equipment 85, 91, 92-3,

101-2, 106, 135
Launch Day 115-16
Moon images 158
problems 130, 134
Galle, Johann 241
gamma radiation 22, 23, 24
Gamow, George 16
Gassendi, Pierre 32
Geiger counters 96
geyser plumes 244
giant stars 34
Gilbert, William 39
Gill, David 25
Giotto probe 74
Global Surveyor spacecraft 186, 188
global warming 140-41, 188
Goddard, Robert H. 98
Gold, Thomas 16
Grand Tour missions 56, 83-5, 90, 105
gravity 25, 33, 43, 47, 157
    artificial 170
    reduced 168-70, 171
gravity assist 81-2, 83-5, 86, 117, 175, 255
gravity-mapping 178-9, 180
Great Wall of China 149
greenhouse effect 140-41, 185, 188, 220
Greenstein, Jesse L. 95-6
Greenwich meridian 76, 78
Guillemin, Victor 257

Hale, George Ellery 26
Hall, Asaph 185
Hammel, Heidi 237, 238-9, 241
Harrison, John 146-7
Hayden Planetarium 251
heavy water 24
heliopause 250
helioseismology 26-7
heliostat 26
helium 17, 21, 22, 139, 199, 205, 217, 232, 242
Herschel, William 183, 222, 230
Hertz, Heinrich 120
Hertzsprung-Russell diagram 14
Hey, James Stanley 48
Hodgson, Richard 26
Hohmann transfer 117
Hooke, Robert 200
Hoyle, Fred 17
Hubble Deep Field 17
Hubble Space Telescope 17, 47, 49, 51, 52, 70, 212, 229, 249, 255
    Europa 209
    Neptune 242, 243
    Pluto 249
    Saturn 215
    Uranus 236, 237

humans
    effect of Sun on 18
    effects on Earth's environment 140-41
    space travel 52, 61-8, 108-10, 168-70, 188, 197
Huygens, Christiaan 25, 216
Huygens probe 92, 93, 95, 194
hydrogen 87, 102, 139, 199, 210, 217, 232
ionized 21

ices 252-4
imaging technology 47, 49-51
inferferometry 101, 132, 233
infrared 23, 51, 101, 233
instrumentation 89-111
intelligence 58, 87, 258-9
International Space Station (ISS) 62, 66-7, 105, 168
iridium 71
IRIS (Infrared Interferometer Spectrometer and Radiometer) 101, 233
Italian Space Agency 123

Jansky, Karl 48
Japanese space research 28, 64, 113, 122, 160, 197
Jet Propulsion Laboratory (JPL) 51, 53, 56, 57, 79, 83, 85, 90, 106
Jewett, David 255
Johnson Space Flight Center 53, 115
Jupiter 68, 199-213, 238
    auroras 52, 200, 205
    future exploration 64, 108, 110
    Galileo flights 40, 51, 52, 56, 61, 85, 132, 203, 205-9
    Great Red Spot 40, 200, 202, 203
    lightning storms 202, 203
    magnetic fields 205-6
    moons 37, 39, 46, 52, 84, 85, 197, 200, 206-13, 253, see also Europa
    Pioneer flights 56, 203
    rings 200, 213
    Voyager flights 49, 61, 75, 83-5, 91, 124, 135, 200-206, 209, 212-13

Kant, Immanuel 13-14, 47
Kennedy, John F. 55-6
Kepler, Johannes 35, 41-3, 47, 69, 163, 170, 174
Kepler satellite 45-6
Kibalchich, Nikolai I. 98
Kohlhase, Charles 75, 83-4
Koronos-F 27
Kuiper Belt 37, 40, 74, 254, 255-7
Kuiper Belt Objects (KBOs) 251, 256
Kuiper, Gerard 185, 187, 230, 255
Kuiper Observatory 236

Lassell, William 230, 241
launch day procedure 115-16
launch systems 54, 90, 98, 107-13
Lee, Pascal 65-7
Leighton, Robert 185
Lemaître, Georges Edouard 16
Lenard, Roger X. 108-10
LeVerrier, Urbain J.J. 175, 237, 241
Ley, Willy 98
life
    creation on Earth 140, 224-7, 253
    extraterrestrial 37, 40, 87, 192-4, 211, 221, 224-7
    future of 258-9
light studies 219-20
longitude 76, 146-7, 156
Lovell, Bernard 48
Lowell, Percival 166, 167, 183, 249
Luna missions 39, 79, 158
Lusser, Robert 98
Luu, Jane 256
Lyttleton, Raymond A. 251

McCord, Thomas 244-5
McKay, Chris 186, 188, 253
Mädler, Johann H. von 76, 183
Magellan spacecraft 51, 177, 178-80
magnetic fields 26, 28, 32-4, 155-6, 205-6, 233
magnetometers 101, 206, 213
Magnum launcher system 113
mapping
    of Earth 39, 145-8
    of solar system 38-40
    see also names of planets
Marconi, Guglielmo 120, 125
Mariner flights 79, 124, 135, 190
Mariner 2 55, 125, 177-8
Mariner 4 55, 86, 187
Mariner 6 190
Mariner 7 190
Mariner 9 190
Mariner 10 55, 81, 167, 181
Mars 55, 68, 163, 183-97
    atmosphere 185-6, 190, 191
    'canals' 183, 187, 190
    communication with Earth 66-7
    future exploration 52, 62-8, 108-10, 168-70, 187, 188, 197
    Global Surveyor (MGS) 186, 188, 191, 196
    life on? 40, 185, 187, 191-5, 221
    mapping 39, 52, 76, 96, 186-7
    Mariner missions 55, 86, 135, 187-90
    Mars Express/Beagle lander 194, 197
    meteorites from 69, 194, 197, 224-7
    moons (Phobos and Deimos) 37, 185, 190

Odyssey spacecraft 197
polar caps 183, 185, 190, 253
Viking missions 191, 192, 195
water evidence 185, 191, 192-5, 196, 197
Maskelyne, Nevil 78
Mauna Kea, Hawaii 48, 51, 236, 239, 241, 243
Maunder Minimum 29
Maxwell, James Clerk 124, 134
Mayor, Michel 45
Mercator, Gerardus 146
Mercury 48, 68, 163, 164-75, 181-2
BepiColombo mission 175, 181-2
craters 166, 167, 173
mapping 39, 52, 164-6, 175
Mariner missions 55, 81, 167-72, 181
MESSENGER mission 173-5, 181-2
polar caps 167
smooth plains 172, 173
temperature 164, 167
Meteor Crater, Arizona 71, 155
meteorites 69, 155, 194, 197, 223, 224-7
meteoroids 69, 166
methane 221, 222, 230, 242, 247, 252
methane ice 244, 251, 252, 253, 255
micrometeor detectors 101
microwaves 23, 122-3
Milky Way 19, 35, 37, 48, 51
mind, and the future 258-9
Mir space station 66, 171, 168
Mira 34, 35
Miranda 234-5
Mohorovicic, Andrija 154
Moon 39, 64, 66-7, 142, 157-61
colour 158
craters 157
and eclipses 19-20, 33
future exploration 52, 62-7, 160
human landings 55-6, 107, 158, 160
robot landings 55, 96, 158
as seen from space 158, 159, 160-61
water possibilities 160
moons in solar system 37, see also names of planets
Murray, Bruce 185

nanosatellites 152
NASA, 46, 52, 53, 56-7, 64, 70, 158-60, 197, see also Deep Space Network; Grand Tour missions; MESSENGER
NASA/ESA joint projects, see Hubble Space Telescope; Next Generation Space Telescope; SOHO
navigation from Earth 126-34
navigation in space 74-9
Near Earth Objects 70-73

NEAR (Near Earth Asteroid Rendezvous) mission 70
Neptune 49-51, 68, 77, 229, 237-47
clouds 49-51, 239, 241-2, 247
dark spots 239, 243
Hubble Space Telescope 239, 242, 243
interior 233, 243, 246, 247
moons 37, 124, 127, 128, 242, 243, 244-5, 253
rings 243
storms 49-51, 239
Voyager flights 79, 83-5, 91, 122, 124, 127-8, 241
winds 40, 243, 247
NERVA (Nuclear Engine for Rocket Vehicle Applications) 109, 111
neutrinos 22-4
neutron spectroscopy 160
Newton, Isaac 14, 25, 43, 46, 47, 147, 174
Next Generation Space Telescope (NGST) 49
Nicholson, Seth 175
Nimbus satellite 91
nitrogen 139, 187, 220, 244
northern lights 32
novae 35
nuclear batteries see radioisotope thermoelectric generators
nuclear electric propulsion (NEP) 108-10, 111
nuclear fusion (Sun) 21-4
nuclear power for spacecraft 108-12

Oberth, Hermann 98
occultations 47-8, 255
O'Dell, Robert C. 45
Odyssey spacecraft 197
Oort Cloud 37, 254, 256
Orbiting Solar Observatory 27
Orion 19
Orion Nebula 45
orreries 78-9
oxygen 139, 156, 188, 210
ozone 212
ozone depletion 141, 151

panspermia theory 87
photography 27, 39, 148, 152
photometers 101
photons 17, 23
photosphere 19-20, 26-7
Piazzi, Giuseppe 70
Picard, Jean 78
Pickering, William H. 249
Pioneer spacecraft/flights 51, 52, 55, 56, 81, 87, 91, 92, 106, 128, 177, 178, 203
planetesimals 15, 241
planets 37, 39-40, 56-7, 68-9, 77
distances between 37-8, 77

elliptical orbits 41-3, 69
mnemonic 69
in other solar systems 45-6
Pluto 39, 40, 68-9, 247-55
Hubble Space Telescope 249, 251, 255
mapping 39, 49
moon (Charon) 37, 249, 250-51, 254, 255
pollution 141
probe instrumentation 102
protons 22, 23
Ptolemy 40, 43, 146, 158
pulsars 43-4, 48

quasars 48
Queloz, Didier 45

radar observation and imaging 46, 48, 79, 151, 152, 167
radiation see electromagnetic radiation
radio equipment 101, 120-25
radio telescopes 43, 44, 46, 47-8, 120, 122, 175
radio waves 23, 120, 243-6
radioisotope thermoelectric generators (RTGs) 83, 91-3, 102, 103-6
radiometers 101, 149, 178, 233, 243
ramjet engine 112
red giants 19, 34
redshift 16
relativity theory 33, 174
remote imaging 20, 147-53
Rheticus, George Joaquim 40
Riccioli, Giovanni 158
Richer, Jean 25
rocket systems 98-9, 107-13
Russell, Henry Norris 14
Russian space research 27, 28, 54, 55, 66, 96, 113, 152, 191, see also Luna; Mir; Sputniks; Venera

Sagan, Carl 48, 72, 178, 180, 197, 226, 257
St. John, Charles 175
San Andreas Fault 150, 151
satellites (artificial) 27-8, 52, 54, 105, 148-53, see also names of spacecraft, satellites (natural), moons.
satellites (natural) see moons
Saturn 49, 68, 199-200, 215-23, 230
Cassini mission 51, 52, 56, 85-6, 125, 216-17, 220
mapping 52
moons 37, 48, 220-23, 230, see also Titan
Pioneer flights 56, 81, 215, 219
radar astronomy 48

rings 40, 124, 215-18, 220
  storms 52, 215
  Voyager flights 83-5, 91, 215, 221
Saturn launcher systems 107
Scarf, Frederick L. 219
Schiaparelli, Giovanni 76, 166, 167, 183
Schwabe, Heinrich 25
Sciama, Dennis W. 17
Seebeck Effect 103
seismic mapping 154
seismology and the Sun 26-7
SETI (Search for Extraterrestrial Intelligence)
Institute 87
SETI@home 87
Shoemaker-Levy 9 205
shooting stars 69
Siemienowicz, Kasimierz 98
Singer, Fred 148
Sirius 59, 76, 200
Skylab 168
Slipher, Vesto 250
SOHO (Solar and Heliospheric Observatory) 20,
27, 28
solar eclipses 19-21, 33
solar flares 26, 28, 29-32
solar ion engines 112
solar maximum and minimum 29
solar power for spacecraft 91-2, 102-5,
112, 177
solar prominences 29
solar radiation 13, 18, 21-4
solar sails 112
solar spectrum, mapping 25
solar storms 18, 20, see also solar flares
solar system models 78-9
solar systems, other 43-6, 51
solar wind 15, 28, 74, 96, 153, 185
solid rocket boosters (SRBs) 107, 113
solid state recorders systems (SSRs) 94-5
Soyuz missions 28, 168
space, Arthur C. Clarke on 58-9
space agencies 53, 54
Space Flight Operation Facility (SFOF) 126-30
space flight planning 74-9
Space Infrared Telescope Facility 46
Space Shuttle 49, 62, 85, 107-13, 115-16,
134, 147, 151
  Launch Day 115-16
  Radar Topography Mission (SHTM) 150
space sickness 168-70
spacecraft 39, 52, 55-7, 89-111, 158, see
also names of craft
Spaceguard 70, 72
spectra 219-20, 233
spectrographs 45, 49, 212
spectroheliograph 26
spectrometers 101, 102, 151, 197, 243

Sputniks 55, 96, 148
stars 14, 19, 34-5, 45
stars with planets (SWP) 144
Steady State theory 16-19
Steel, Duncan 252-4
steering and navigation 126-34
Stern, Alan 256, 257
Stoney, George I. 185
stratospheric dust 71
Stuhlinger, Ernst 98
Sun 11-28
  death of 34-5
  distance from Earth 25
  effect on human life 18
  see also solar...
sunspots 25-6, 27, 29, 38, 39
supergiants 19
supernovae 16, 35, 38, 48
synthetic aperture radar (SAR) 151, 178

telemetry and computing 120-35
telescopes 17, 25-6, 41, 48-51
very large 48, 51, 120, 122
see also radio telescopes; optical telescopes
Terra satellite 149
terraforming 188
thermal control 92-3
Thermoelectric Outer Planets Spacecraft (TOPS)
56, 90-91, 105
tides 157, 244
time 58, 146, 259
Titan 48, 93, 124, 194, 220-22, 253
Titan launcher systems 107, 111, 116
Toftoy, Holger 98
Tombaugh, Clyde W. 249-50
TRACE (Transition Region and Coronal Explorer)
27-8
tracking 54, 90, 125, 131-4
trajectory planning 116-20
trans-Neptunian objects (TNOs) 254
Transit Earth satellites 91, 105
Triton (moon of Neptune) see Neptune, moons
Tsiolkovsky, Konstantin E. 95, 98
Tycho Brahe 35, 40-41, 43

ultraviolet light 18, 20, 23, 27, 102, 141
Ulysses probe 27, 28
universe, future of 258-9
universes, many others 17-19
Uranus 37, 68, 229-37, 238
  aurorae 233
  clouds 232, 236
  future exploration 109
  Hubble Space Telescope 229, 232,
  236, 237
  magnetic field 233
  moons 37, 128, 230, 232, 233-6, 237

rings 232, 236, 237
seasonal changes 232, 238
tilting 40, 128, 230, 233, 238
Voyager flights 83-5, 91, 127-8, 132,
135, 230, 233, 234, 236, 237, 238

Van Allen belts 96, 148
Vanguard satellites 105
VASIMR (Variable Specific Impulse
Magnetoplasma Rocket) 111, 112
Vega 58
velocity 82-3, 125, 127
Venera spacecraft 93, 177, 178
Venus 39, 40, 48, 55, 68, 163, 175-82
  clouds 175, 178, 180
  craters 179
  gravity assists 85, 117, 175
  gravity-mapping 178-9, 180
  Magellan mission 178-80, 182
  mapping 39, 52, 179-80, 182
  Mariner missions 55, 81, 125,
  134, 177-8
  retrograde rotation 177
Viking spacecraft 107, 185, 191, 192,
195, 224
visible light 23, 27, 30
volcanoes 38, 147, 154
  on Jupiter's moons 52, 206-8, 210, 253
  on Mars 186-7
  on Venus 178, 179-80
Voyager 1 and 2 52, 56-7, 101, 106, 107,
253, see also Jupiter; Neptune; Saturn; Uranus
Vulcanoids 174

Waldseemüller, Martin 146
Wegener, Alfred 154
weightlessness 64, 168-70
WF/PC (Wiffpick) (HST Wide Field and
Planetary Camera-2) 49, 70, 215, 237, 242
Wilcke, Johan 98
Wildt, Rupert 219
Wolf, Rudolf 26
Wolszczan, Alex 43-4
wormholes 112

x-rays 23, 27, 30

Yokoh 28

Zeeman, Pieter 47
zodiacal cloud 69
Zucchi, Nicolas 200

Further Reading

Stardust: Supernovae and Life; The Cosmic Connection, John and Mary Gribbin (Yale University Press, 2000). Relatively easy reading on solar physics, intermixed with the seductive theme that we are the stuff of stars.

The New Solar System, J. Kelly Beatty and Carolyn Collins Petersen (eds.) (Cambridge University Press, 1999). Extensive, expert, and rather technical information on the bodies of the solar system.

Asteroids, Curtis Peebles (Smithsonian Institution Press, 2001). The story from Chicxulub to the NEAR spacecraft's rendezvous with Eros, with extended commentary on that irresistible question – what if an asteroid were to threaten Earth?

Flight to Mercury, Bruce Murray and Eric Burgess (Columbia University Press, 1977). Includes the politics and economics of spaceflight, and more than one hundred high-resolution photographs of Mercury's surface.

Venus Revealed, David Harry Grinspoon (Perseus, 1998). Beautifully crafted read about an intriguing planet, graced with a strong personal voice.

The Earth in Context: a Guide to the Solar System, David Harland (Springer, 2001). How Earth's geological history was discovered, and how it compares with the other planets.

Mapping Mars, Oliver Morton (Fourth Estate (UK), Picador (US), 2002). A cartographic and geographical tour-de-force, from spacecraft exploration to what Martian dirt looks like and how we'd survive if we ever had to walk on it.

Mars: The Lure of the Red Planet, William Sheehan and Stephen James O'Meara (Prometheus Books, 2001). A tribute to Earth-based observations of the Red Planet, replete with notes from famous astronomers and tips on how to observe Mars with your own telescope.

Jupiter: the Giant Planet, Reta Beebe. (Smithsonian Institution Press, 1997). A concise classic still, the updated 2nd edition includes the Shoemaker-Levy 9 collision and some of the Galileo spacecraft discoveries.

Lifting Titan's Veil: Exploring the Giant Moon of Saturn, Ralph Lorenz, Jacqueline Mitton (Cambridge University Press, 2002). Titan, larger than the planets Mercury and Pluto and the second largest moon in the solar system, will be visited by the Huygens space probe in 2004. This is its story.

Neptune: The Planet, Rings and Satellites, Ellis Miner and Randii Wessen (Springer, 2001). Comparisons with Jupiter are sprinkled through the book, which ends with informational summaries on all of the planets.

Pluto and Charon: Ice World on the Ragged Edge of the Solar System, S. Alan Stern and Jacqueline Mitton (Wiley, 1999). Not much is known about this far-out, icy pair, but what is know, is here, from Tombaugh's discovery to newer information gleaned from spacecraft and earth-based telescopy.

Rockets into Space, Frank Winter (Harvard University Press, 1990). A concise, readable review of rocketry with a preview of futuristic space propulsion proposals.

Entering Space: Creating a Spacefaring Civilization, Robert Zubrin. (J. P. Tarcher, 2000). The irrepressible author of *The Case for Mars* explains why it's only natural that the human species should continue its migrations into space.

And more...

Physical data on the planets are updated regularly at
http://solarsystem.nasa.gov/features/planets/ planet_profiles.html
See http://www.thesolarsystem.org for a longer list of books and web links.

Picture Credits

p2 Moonruner Design; p10 Moonrunner Design; p12 courtesy of NASA/JPL/Caltech /SOHO/Extreme Ultraviolet Imaging Telescope (EIT) consortium); p14 Patrick Mulrey; p15 Douglas Lin and Geoff Bryden, University of California at Santa Cruz; p19 Moonrunner Design; p22 Patrick Mulrey; p23 Patrick Mulrey; p26 SPL; p28 Nico Cheetham; p29 Courtesy of SOHO/EIT consortium. (SOHO is a project of international cooperation between ESA and NASA); p30 NASA/JISAS/SPL; p31 NASA/TRACE; p36 Moonrunner Design; p40 SPL; p41 SPL; p42 SPL; p43 SPL/Dr Jeremy Burgess; p44 SPL; p48 SPL; p49 NASA/STScl; p49 NASA/STScl; p50 NASA; p60 Moonrunner Design; p63 NASA; p67 Julian Baum/SPL; p70 NASA (top); NASA/Science Magazine (bottom); p73 Scott Manley and Duncan Steel; p74 ESA/SPL; p78 SPL; p80 Moonrunner Design; p83 Patrick Mulrey; p84 Patrick Mulrey; p85 Patrick Mulrey; p88 Moonrunner Design; p91 Roy Flooks; p92 Roy Flooks; p93 Roy Flooks; p97 NASA; p99 NASA; p103 Roy Flooks; p107 Roy Flooks; p109 Roy Flooks; p110 NASA; p111 Roy Flooks; p112 Roy Flooks; p113 Roy Flooks; p114 Moonrunner Design; p117 Patrick Mulrey; p121 David Nunuk/SPL; p123 Patrick Mulrey; p124 Patrick Mulrey; p127 NASA; p136 Moonrunner Design; p138 F. Hasler, M. Jentoft-Nilsen, H. Pierce, K. Palaniappan, and M. Manyin. NASA Goddard Lab for Atmospheres/Data from National Oceanic and Atmospheric Administration (NOAA); p139 NASA; p143 Johnson Space Centre/NASA; p147 NASA; p149 (bottom left) NASA/GSFC/METI /ERSDAC/JAROS, and U.S./Japan ASTER Science Team; p149 (bottom right) NASA/GSFC/METI /ERSDAC/JAROS, and U.S./Japan ASTER Science Team; p149 (top) NASA/GSFC/METI /ERSDAC/JAROS, and U.S./Japan ASTER Science Team; p150 NASA/JPL/ NIMA; p152 NASA; p153 Moonrunner Design; p155 Moonrunner Design; p159 NASA; p161 NASA; p162 Moonrunner Design; p165 NASA; p166 Moonrunner Design; p169 Moonrunner Design; p171 ESA/SPL; p173 NASA/JPL/Northwestern University; p176 NASA; p177 Moonrunner Design; p178 NASA; p179 Novosti/SPL; p180 NASA; p183 SPL; p184 NASA; p185 Moonrunner Design; p186 NASA/JPL /Malin Space Science Systems; p187 NASA; p188 Paul Bourke; p189 Paul Bourke; p190-192 NASA; p193 NASA/JPL/Malin Space Science Systems; p195 NASA; p196 NASA/JPL/ Malin Space Science Systems; p197 NASA/JPL/University of Arizona/Los Alamos National Laboratory; p198 Moonrunner Design; p200 NASA/Hubble Heritage Team(STScl/AURA)/ESA/ John Clarke (University of Michigan); p201 NASA/JPL/University of Arizona; p202 NASA; p203 NASA; p204 Moonrunner Design; p207 NASA; p208 (left) NASA; p208 (centre) NASA; p208 (left) NASA/JPL/DLR (German Aerospace Center) ; p209 (left) NASA; p209 (right) F. Hasler, M. Jentoft-Nilsen, H. Pierce, K. Palaniappan, and M. Manyin/NASA Goddard Lab for Atmospheres/Data from National Oceanic and Atmospheric Administration (NOAA); p210 (top) NASA; p210 (bottom) NASA; p211(right) Moonrunner Design; p211 (left) NASA; p212 NASA/JPL/ University of Arizona; p213 NASA; p213 (bottom) NASA; p214 NASA; p216-217 (all images) NASA/The Hubble Heritage Team (STScl/AURA) Acknowledgment: R.G. French (Wellesley College), J. Cuzzi (NASA/Ames), L. Dones (SwRI), and J. Lissauer (NASA/Ames); p218 Moonrunner Design ; p220 NASA; p221 NASA; p222 (left) NASA; p222 (top right) NASA; p222 (bottom right) NASA; p223 (left) NASA; p223 (right) NASA; p225 NASA; p226 NASA ; p227 NASA; p228 Moonrunner Design; p231 NASA; p232 Moonrunner Design; p234 (top) NASA; p234 (bottom) NASA; p235 NASA; p236 Kenneth Seidelmann, U.S. Naval Observatory/NASA; p240 NASA; p242 NASA; p245 NASA; p246 Moonrunner Design; p248 Dr. R. Albrecht/ESA/ESO Space Telescope European Coordinating Facility/NASA; p249 Alan Stern (Southwest Research Institute) /Marc Buie (Lowell Observatory) /NASA/ESA ; p250 Moonrunner Design; p252 NASA; p256 Moonrunner Design.

Acknowledgements

Above all I wish to express my thanks to Ellis Miner and Charley Kohlhase, who encouraged and mentored me through the final stages of this book. Charley Kohlhase led the mission design teams for many Jet Propulsion Laboratory projects over a period of 35 years, and now consults on the Mars programme. Ellis Miner is also with JPL, where he is the science manager for the Cassini Mission, scheduled to arrive at Saturn on July 1, 2004.

Grateful thanks are also due to Philip Campbell for putting me in touch with the right people at the right time – Peter Tallack in particular – to make the book a reality. I am also deeply indebted to Heidi Hammel of the Space Science Institute based in Boulder, Colorado, and many others who cared enough about the communication of science to put aside their professional work to read other sections of the book. These include a number of other eminent scientists — Michael Abrams (Jet Propulsion Laboratory), K. S. Balasubramaniam (National Solar Observatory), Don Brownlow (University of Washington), Clark Chapman (Southwest Research Institute), Heiko Hecht (MIT Man-Vehicle Lab), Chris McKay (NASA Ames Research Center), Renu Malhotra (University of Arizona), Joel Parker (Southwest Research Institute), Suzanne Smrekar (Jet Propulsion Laboratory), Jill Tarter (SETI Institute), Fred Taylor (Oxford University); and Wes Traub (Harvard-Smithsonian Center for Astrophysics).

My reviewers advised me only on those subjects in which they are expert, and since the book continued to develop somewhat after their reviews, are not responsible for any errors that may have occurred. On the other hand, they contributed an edge of excellence that would not have existed otherwise.

David Morrison and Heidi Hammel also contributed two of the eleven signed essays that are included in this volume. The other authors are, in alphabetical order, Donald Brownlee, Arthur C. Clarke, David Darling, Sarah Dunkin, Freeman Dyson, Donald Gray, Pascal Lee and Stephen Braham, Roger Lenard, and Duncan Steel. I am deeply honoured that their writings appear beside my own.

A number of other individuals — scientists, educators, and communicators — helped in diverse and valuable ways. They included Dee Foster, Luke Dones, John Guest, Gerhard Hahn, Martha Heil, Floyd Herbert, Steve Maran, Paul Murdin, Leik Myrabo, Jay Pasachoff, Tim Radford, Sean Solomon, Alan Stern, Marguerite Syvertson, and Guy Webster. I was also encouraged greatly by the warm camaraderie of fellow science writers on both sides of the Atlantic.

Finally, I wish to offer my grateful thanks to the other members of the Solar System team, whose considerable talents were, at the end of the day, more crucial than any other in bringing the book successfully to press. These include my editor Nic Cheetham, artist Malcolm Godwin, and the talented backstage crew who provided everything else, from copy-reading the text to choosing the paper on which it is printed.